Measuring Slipperiness

Measuring Slipperiness

Human Locomotion and Surface Factors

EDITED BY

Wen-Ruey Chang, Theodore K. Courtney, Raoul Grönqvist and Mark Redfern

CRC Press
Taylor & Francis Group
Boca Raton London New York

CRC Press is an imprint of the
Taylor & Francis Group, an **informa** business

A TAYLOR & FRANCIS BOOK

First published 2003 by Taylor & Francis

Published 2019 by CRC Press
Taylor & Francis Group
6000 Broken Sound Parkway NW, Suite 300
Boca Raton, FL 33487-2742

© 2003 by Taylor & Francis Group, LLC
CRC Press is an imprint of Taylor & Francis Group, an Informa business

First issued in paperback 2019

No claim to original U.S. Government works

ISBN 13: 978-0-367-45463-0 (pbk)
ISBN 13: 978-0-415-29828-5 (hbk)

Visit the Taylor & Francis Web site at
http://www.taylorandfrancis.com

and the CRC Press Web site at
http://www.crcpress.com

Typeset in Times by Elite Typesetting Techniques Ltd.

Every effort has been made to ensure that the advice and information in this book is true and accurate at the time of going to press. However, neither the publisher nor the authors can accept any legal responsibility or liability for any errors or omissions that may be made. In the case of drug administration, any medical procedure or the use of technical equipment mentioned within this book, you are strongly advised to contact the manufacturer's guidelines.

British Library Cataloguing in Publication Data
A catalogue record for this book is available from the British Library

Library of Congress Cataloging in Publication Data

Measuring slipperiness: human locomotion and surface factors/edited by Wen-Ruey Chang and Theodore K. Courtney; associate editors, Raoul Gronqvist, Mark Redfern.
   p. cm.
Includes bibliographical references and index.
ISBN 0-415-29828-8 (hb.)
1. Industrial safety–United States. 2. Human locomotion. 3. Surfaces (Technology) 4. Friction–Measurement. 5. Falls (Accidents)–United States–Prevention. I. Chang, Wen-Ruey. II. Courtney, Theodore K. III. Gronqvist, Raoul.

T55 .M395 2003
620.8'6–dc21   2002032077

Contents

Preface

World-wide, falls represent the third leading cause of disability after the systemic conditions of depression and anaemia. Falls are also the second leading global cause, after motor vehicle collisions, of accidental death. In developed countries, slips, trips and falls (STF) contribute between 20 and 40% of disabling workplace injuries. The precise contribution of slipping and slipperiness to this burden are not completely understood. In fact, one of the most intractable problems in STF research and prevention is that of defining and, in particular, measuring slipperiness.

Various measurements of slipperiness have historically played an important role in advancing the understanding of slip and fall mechanisms and in the development of preventive interventions. The most common and perhaps oldest methods have involved the measurement of friction. However, slipperiness is a complex concept involving aspects of biomechanics, tribology, neurophysiology and human cognition. Subjective or human-centred assessments, roughness measurements of floor and shoe surfaces, and biomechanical measurements might be good alternatives to friction as measures of slipperiness. Even so, significant, often controversial, questions remain regarding the conceptualisation, definition, and measurement of slipperiness.

On 27–28 July 2000, the Liberty Mutual Research Center for Safety and Health (LMRC) hosted its second two-day international symposium on science and research methods in Hopkinton, Massachusetts. The programme, organised by researchers at the LMRC, the Finnish Institute of Occupational Health (FIOH), and the University of Pittsburgh, sought to develop an international perspective on methodological issues in slipperiness measurement.

Symposium participants were researchers noted for their work and record of scholarly accomplishment in various aspects of STF and slipperiness measurement. The nearly 30 participants brought perspectives from Australia, France, Finland, Germany, Korea, Sweden, Thailand, the United Kingdom and the United States. Among the participants were representatives from FIOH, the French Institut National de Recherche et de Sécurité—INRS, the UK Health and Safety Laboratory, the US National Institute for Occupational Safety and Health and premier international universities.

Five months prior to the conference participants were organised on a technical affinity/self-selection basis into collegial, issue-focused working groups, including concepts and definitions, slipperiness and occupational injury, and biomechanical, human-centred, surface roughness, and friction-based measurement approaches. The working groups laboured for several months to produce manuscripts for extensive discussion and critique at the symposium. Each group was responsible for presenting the state of the science in its area and to provide guidance on important issues and improvements for future studies.

At the meeting, each working group presented its draft manuscript and was then rebutted by another working group assigned as its technical 'opponent'. The participants as a whole then critically reviewed the subject area and discussed limitations and recommendations for improvement. Each manuscript also received reviews from two or more 'external' reviewers who were not participants in the Measurement of Slipperiness programme. Finally, the working groups returned to their manuscripts to make the suggested revisions and improvements. The collective result of these collegial labours appears in Chapters 1–7 of the book (having appeared previously in a special issue of *Ergonomics*—Vol. 44, No. 13).

Another facet of the Hopkinton Conference on the Measurement of Slipperiness involved surveys of individual participants and working groups and moderated discussions of the whole. These surveys and discussions focused on key issues such as the adequacy of current approaches, areas of potential technical consensus, criteria for evaluating a slipperiness system, and future research needs and directions. The data from these discussions were analysed and synthesised into an assessment of the state of the art and a conceptual framework of directions for future studies that form the book's concluding Chapter 8.

Students, professionals and researchers in ergonomics, health and safety, gait biomechanics, human locomotion, and wear and lubrication sciences will find that *Measuring Slipperiness...* provides a valuable, state-of-the-art reference and supplementary text. Flooring and floor product manufacturers, walking surface designers, footwear manufacturers, and standards organisations should also find the book informative in addressing some of their concerns.

The programme enjoyed a diversity of perspectives (clinical, engineering, psychological, physiological and epidemiological to name a few) which we believe to be both a strength of the work as well as a reason underlying the vigorous debate on many points. It is our hope that this book will lay a foundation from which greater inroads in slipperiness measurement and ultimately STF prevention can be achieved.

WEN-RUEY CHANG
THEODORE K. COURTNEY
RAOUL GRÖNQVIST AND
MARK REDFERN
Hopkinton, Massachusetts, USA
Helsinki, Finland,
Pittsburgh, Pennsylvania, USA

ix

Acknowledgements

The editors gratefully acknowledge the support of the many participating international institutions and thank our colleagues in STF research, whether approaching the problem from the clinical, engineering, psychological, physiological or epidemiological perspective, for their contributions to the project. We offer special thanks to those pioneers in the field such as Lennart Strandberg, Håkan Lanshammar, Robert Brungraber, and Derek Manning who 'came out of retirement' in a figurative or literal sense to share their vast experience with measuring slipperiness with the rest of us. We are indebted to our colleague, Tom Leamon, himself a pioneer in STF research and Director of the Liberty Mutual Research Center for Safety and Health, both for initially suggesting slipperiness as the theme for a methodological symposium and for his unswerving support throughout the project.

We thank General Editor Robert Stammers for his helpful criticisms, encouragement in the preparation of the original special issue of *Ergonomics*, and the thoughtful Foreword he contributed to *Measuring Slipperiness*.... Thanks also to our publisher, Taylor & Francis, Ltd. (especially Imran Mirza at Journals and Tony Moore and Sarah Kramer at Books) who kept us to schedule and handled the challenges we threw them with grace, warmth and industry.

Many thanks to our fellow scientists at the Liberty Mutual Research Center for Safety and Health, the Finnish Institute of Occupational Health, and the University of Pittsburgh for their contributions in review and advice. The staff of the Liberty Mutual Research Center were instrumental in making the programme a success in terms of planning, facilities and logistics support (especially Ed Correa, Susan Flannery, Janet Healy, Julie Twomey and Kathy Whittles) and proofing and indexing (especially Margaret Rothwell). We acknowledge and thank Milja Ahola for her assistance with artwork in Chapter 1 which was adapted for the cover of *Measuring Slipperiness*. Finally, we would each like to thank our families who encouraged us when we were discouraged and bore our absences (of body and/or mind) nobly during the intensive phases of the project.

Foreword

The basis of this book was a special issue of the journal *Ergonomics*. As editor of that journal, I have to confess to some surprise when I was first approached with the idea of a special issue on the topic of the *Measurement of Slipperiness*. The surprise was because this was not a topic that I had ever seen as a critical one before. My reaction changed when I read the detailed proposal. It was clear that slipperiness is not a narrow and highly specialised subject, with limited relevance to ergonomics. On the contrary, it clearly emerges as a central topic in the important field of slips, trips and falls (STF). In turn, this latter area of human misfortune carries with it a heavy cost in terms of injury and death.

The editors of *Measuring Slipperiness*... are to be congratulated on having the foresight to see the central importance of the measurement of slipperiness. More than this they went about producing a state-of-the-art review in a very effective manner. The chapters of this book have their origin in the second "Hopkinton" symposium on science and research methods held at the Liberty Mutual Research Center.

In organising the symposium, the editors did not adopt the more usual approach of calling on individual experts to produce review papers and then putting these papers together with only limited integration. Instead, the organisers formed collegial groups of researchers who collected information and then wrote the initial drafts of papers on the range of topics that had been identified as covering the key issues. These papers were then presented at the symposium. Reviews and comments on each paper were obtained from the other groups and from referees and the papers were then subjected to revision and editorial work.

The result is a set of chapters that can truly be said to be comprehensive statements on the issues. The collegial nature of the work is reflected in the fact that the number of authors of the first seven chapters of this book averages over seven. The last, overview, chapter has been specially produced for this book.

Another important feature of this volume is the international make-up of both the editors and the authors. This leads to a pooling of important resources and data, and reflects the international nature of ergonomics research.

For ergonomics, this book should serve as a useful reminder of the nature of the discipline. The *multi-disciplinary* nature of the sciences that are drawn upon in examining slipperiness measurement is very typical of ergonomics problems. However, a recognition of the need for an *interdisciplinary* approach to analysing a problem and formulating solutions is what marks out ergonomics as an integrating discipline. In this field of slipperiness, the range of underlying sciences is as wide as can be found in any ergonomics problem. From physics and engineering through to the cognitive psychology of situation awareness is indeed a broad array of approaches.

It is clear that this book is a useful starting point for the understanding of the many aspects of the STF problem. The particular nature of STFs cries out for solutions that will help us not only understand the phenomenon of human interaction with slippery walking surfaces, but also develop preventive design solutions to ameliorate this universal problem. Those solutions in themselves will be multi-faceted. They will have to do with bringing about change in the nature of floor surfaces, footwear, other forms of locomotion support, the education and training of individuals and a variety of other topics.

Whilst the book cannot be said to offer any complete solutions at present, it clearly defines the field, serves as a repository of the accumulated knowledge, sets the agenda for future research and sets the standard for the quality of research that needs to follow. As such it is a very important contribution to the knowledge of an important field. Just as importantly, the approach adopted for getting researchers to collaborate serves as an exemplar case for how the multi-faceted problems of ergonomics can be tackled in the future.

Professor Rob Stammers
University of Leicester, UK
General Editor, *Ergonomics*

CHAPTER 1

Measurement of slipperiness: fundamental concepts and definitions

Raoul Grönqvist¶*, Wen-Ruey Chang†, Theodore K. Courtney†,
Tom B. Leamon†, Mark S. Redfern‡ and Lennart Strandberg§

¶Finnish Institute of Occupational Health, Department of Physics,
Topeliuksenkatu 41 FIN-00250 Helsinki, Finland

†Liberty Mutual Research Centre for Safety and Health, Hopkinton, MA 01748,
USA

‡Departments of Otolaryngology and Bioengineering, University of Pittsburgh,
Pittsburgh, PA 15213, USA

§ITN Department of Science and Technology, Linköping University, SE-60174
Norrköping, Sweden

Keywords: Concepts of slipperiness; Slipping hazards; Safety criteria; Definition
of terms

The main objective of this paper is to give an overview of basic concepts and
definitions of terms related to the 'measurement of slipperiness' from the onset of
a foot slide to a gradual loss of balance and a fall. Other unforeseen events prior
to falls (e.g. tripping) are sparingly dealt with. The measurement of slipperiness
may simply comprise an estimation of slipping hazard exposures that initiate the
chain of events ultimately causing an injury. However, there is also a need to
consider the human capacity to anticipate slipperiness and adapt to unsafe
environments for avoiding a loss of balance and an injury. Biomechanical and
human-centred measurements may be utilised for such an approach, including an
evaluation of relevant safety criteria for slip/fall avoidance and procedures for
validation of slip test devices. Mechanical slip testing approaches have been
readily utilized to measure slipperiness in terms of friction or slip resistance but
with conflicting outcomes. An improved understanding of the measurement of
slipperiness paradigm seems to involve an integration of the methodologies used
in several disciplines, among others, injury epidemiology, psychophysics,
biomechanics, motor control, materials science and tribology.

1. Introduction

The main research question in the prevention of slip- and fall-related injuries is
comprised of two parts. First, how to measure the exposure to slipping hazards and,
second, how to evaluate the risk of injury? This paper's focus is on the first part of
the question. The term *slipperiness* is defined here as 'conditions underfoot which
may interfere with human beings, causing a foot slide that may result in injury or
harmful loading of body tissues due to a sudden release of energy'. There are, as yet,
no unambiguous slipperiness measurement methodologies and no generally accepted

---

*Author for correspondence. e-mail: raoul.gronqvist@occuphealth.fi safety criteria or safety thresholds for estimating slipping hazard exposures (i.e. conditions with the potential of causing a loss of balance and a fall-related injury or other adverse effects). It is necessary to continuously re-evaluate the output relevant of any measurement of slipperiness methodology, whether based on a biomechanical experiment, a human-centred subjective assessment or a mechanical slip test. Furthermore, there are no undisputed methods to estimate slipping and falling risks, either in terms of the probability and severity of a loss of balance that can lead to a fall, or in terms of the probability and severity of a subsequent injury and disability (transient or permanent). As a result, the number and severity of fall- and overexertion-related injuries caused by slipping remain prodigious among the working population and among the general public, being a concern especially for the elderly population (Courtney *et al.* 2001, Grönqvist *et al.* 2001).

The multitude and complexity of extrinsic, intrinsic and mixed risk factors for slipping and falling, in fact, suggest that the injuries caused by slips are not the result of trivial incidents with simple prevention strategies (Strandberg 1985, Leamon and Murphy 1995, Redfern and Bloswick 1997, Grönqvist 1999, Leclercq 1999). On the contrary, even keeping track of the sequence of events following the initial perturbation (foot slide, loss of balance, or vice versa) is difficult. They are not fully understood due to their complexity and unexpected progression. However, several underlying perceptual, cognitive, psychological, biomechanical, motor control, and tribophysical mechanisms seem to determine the onset and outcome of slipping incidents, including the ways in which the human body reacts and adapts to a perturbation of balance and threat of falling (Chang *et al.* 2001a, b, c, Grönqvist *et al.* 2001, Redfern *et al.* 2001).

The main objective of this paper is to give an overview of the basic concepts related to the 'measurement of slipperiness' with reference to a simple conceptual framework. An attempt is made to define some key terms in the measurement of slipperiness processes beginning with the onset of a foot slide, followed by a gradual loss of balance and a fall. Other unforeseen events to falls, such as tripping, are dealt with sparingly.

2. Unintentional injuries caused by slipping

2.1. *Traumatic injuries*

Foot slippage is the most frequent unforeseen event triggering falls on the same level and may also cause falls to a lower level (Andersson and Lagerlöf 1983, Courtney *et al.* 2001). A fall to a lower level was in fact the major outcome in a survey on serious occupational accidents caused by slipping (Grönqvist and Roine 1993). Grönqvist (1999: 352) defined *slipping* as 'a sudden loss of grip, often in the presence of liquid or solid contaminants and resulting in sliding of the foot on a surface due to a lower coefficient of friction than that required for the momentary activity'. However, a number of other extrinsic and intrinsic risk factors for slips and falls may contribute to the injury events. These factors may be related to, among others, insufficient lighting, poor housekeeping, ageing, vestibular disease, peripheral neuromuscular dysfunction, diabetes, osteoporosis, alcohol intake, and use of anti-anxiety drugs (cf. Grönqvist 1999).

Manning *et al.* (1984, 1988) used the term underfoot accident to describe fall-related and other unintentional injuries where the first unforeseen event was an interaction between the victim's foot and the substrate. Under this definition also other first unforeseen events than slipping were included, for instance tripping, stumbling, missed footing, twisted foot/ankle, trod on air, and collapsed or moved surface. Most often they lead to contact injuries due to impacts and resulted in sprains, contusions, and fractures of the ankle, knee, hip, thigh, wrist, or the lumbar spine (Manning and Shannon 1981, Manning *et al.* 1984, 1988).

A slip- and fall-related injury tends to be a very rapid event lasting less than a second until, for instance, an outstretched hand or the pelvis impacts with the ground. Hsiao and Robinovitch (1998) studied common protective movements associated with falls from standing height and found that a fall was more than twice as likely to occur after anterior translations of the feet (backward falls involving pelvic impact), when compared to posterior or lateral translations (forward falls). This backward fall mechanism appears to be similar to a common injury mechanism due to slipping after heel contact in walking (Strandberg 1985, Grönqvist 1999, Leclercq 1999).

2.2. *Harmful loading of the body*
A foot slide may lead to a balance recovery, if the response time to adjust gait is short enough to avoid a fall. However, it could still cause overexertion and non-contact injuries due to sudden, strenuous movements and unexpected loading of body parts, such as the low back (Strandberg 1985, Stobbe and Plummer 1988). Body movements made to restore balance and to prevent a fall have been found to create substantial muscle forces and harmful loading on the spine (Lavender *et al.* 1988).

Back pain has been reported as a common outcome of slip and trip incidents. However, the role of slipperiness, for example, is not yet understood in the aetiology of back pain. Manning and Shannon (1981) and Manning *et al.* (1984) found that slipping and other underfoot events caused two-thirds of all cases of accidental low-back pain in a car factory. Murphy and Courtney (2000) reported that low-back pain was often (21% of low-back pain claims cost) related to discrete antecedents such as falls.

Extensive friction utilisation during manual materials handling tasks causes a potential risk of a foot slide in the shoe/floor interface (Grieve 1983). There is an urgent need to better understand the role of slipping as a contributing factor to low-back pain and injuries during heavy physical work. A review by Hoogendoorn *et al.* (1999) showed moderate evidence that heavy physical work is a risk factor for back pain, and strong evidence that manual materials handling (lifting, carrying, and holding loads) as well as bending and twisting were risk factors for back pain. For example, pushing and pulling has been mentioned as potential setting for slips and falls as well as for musculoskeletal disorders (Hoozemans *et al.* 1998, Ciriello *et al.* 2001).

2.3. *Slipping versus tripping*
The common factor in all the above described unintentional injuries is that the victim's postural control mechanisms fail to restore balance and stability before injury occurs. Slipping injuries in particular, but also tripping injuries, can be related to the frictional characteristics of shoe soles, floor surfaces and contaminants. Probably tripping and stumbling have more to do with unstable and erratic foot trajectories on uneven or discontinuous surfaces, or with obstacles on the floor, rather than high friction or large variations in the friction coefficient between adjacent floor areas. A trip may occur when the swing phase of the foot is suddenly interrupted (Redfern and Bloswick 1997). The swinging foot may fail to adequately clear the ground or it may encounter an obstacle or an irregularity on the ground surface (Winter 1991). The most common protective movement outcome in response to an

early swing perturbation, according to Eng *et al.* (1994), was an elevating strategy (i.e. flexion of the swing limb). In response to a late swing perturbation, a rapid lowering strategy of the swing limb to the ground and a shortening of the step length was accomplished (Eng *et al.* 1994). The late swing perturbation posed a greater threat for a fall, because the body mass was already anterior to the stance foot.

3. Basic concepts for the measurement of slipperiness

3.1. *Gait and the risk of slipping*

Carlsöö (1962: 271) describes gait as follows: 'In the human gait the equilibrium is lost and regained with every step; lost with the take-off of the propelling foot when the body's centre of gravity momentarily lies beyond the anterior border of the supporting surface, and regained as soon as the swinging leg is extended forward and its heel touches the ground'. Timing and placement of successive steps must be continuously adjusted in order to maintain dynamic equilibrium of the body during walking (Nashner 1980). This inherent instability of upright posture is exploited to assist in propelling the body forward by a forward orientation of the body relative to the feet.

The foot trajectory needs to be controlled during the gait cycle for safe ground clearance and a gentle heel or toe landing (Winter 1991). To allow adequate ground clearance during the swing phase, the leg is flexed at the knee and dorsiflexed at the foot and toes. A gentle heel landing, on the contrary, reduces collision-forces in the shoe/floor interface during weight acceptance (Cappozzo 1991). This, in turn, appears to be important for minimising hydrodynamic load support and maximising friction and slip resistance in the presence of contaminants such as water, oil, or snow (Grönqvist 1999).

From the safety point of view, there are two critical gait phases in level walking (Perkins 1978, Strandberg and Lanshammar 1981). First, the early heel contact — when only the rear part of the heel region is in contact with the ground and, second, the moment of toe-off — when there is only sole forepart contact. The heel contact phase is considered to be more challenging for stability and more hazardous from the slipping point of view than the toe-off phase, because the forward momentum maintains the body weight on the leading foot causing a forward slide of the foot (Redfern *et al.* 2001).

The adaptation to slipperiness (on a dry low-friction surface) seems to involve a maximisation of stability through postural changes during early stance and mid-stance (Llewellyn and Nevola 1992). The body's centre of mass is moved forward closer to the base of support, knee flexion is increased facilitating a softer heel (sole) landing, and ankle plantarflexion is greater than during a normal initial contact phase, thus reducing the heel (sole) contact angle with respect to the floor. Hence, the shoe/floor contact area appears to increase during heel touch-down. Due to lower shear forces available in the shoe/floor interface, the ground reaction force profiles are altered for minimising frictional utilisation. These combined effects of force and postural changes seem to aim at reducing the vertical acceleration and forward velocity of the body (Llewellyn and Nevola 1992).

During safe locomotion, a dynamic interplay needs to exist between the sensory systems (vision, vestibular organ, and proprioception) that control posture, gait and balance (Grönqvist *et al.* 2001, Redfern *et al.* 2001), and the frictional surface phenomena between shoes and walkways (Chang *et al.* 2001a, b). A prompt and accurate human response to this flow of information appears to be a necessary propensity for fall avoidance.

3.2. *A conceptual framework*

A conceptual framework for slipping and the measurement of slipperiness processes is presented in figure 1. This scenario exemplifies the mechanics of a foot slide during habitual gait, starting with heel contact and leading to a subsequent backward fall. A fall on the same level (on walkways, pavements, etc.) during an early stance perturbation probably represents the most common unintentional slip- and fall-incident. Slipping might also lead to an injurious fall to a lower level (stairway falls, falls from a roof, etc.) or to injury due to an overexertion of body parts even if balance is recovered (low-back injury, sprained ankle, etc.). The type and severity of these injuries is influenced by the presence or absence of a variety of other concurrent exposures with the loss of balance, such as dangerous hand-held tools or vehicles in motion (Grönqvist and Roine 1993).

The magnitude for a given risk exposure to slipping and falling hazards depends on the exposed person's health status, anthropometry, perception and cognition of the hazards and the ability to control one's posture and regain perturbed balance (Strandberg 1985, Tisserand 1985, Redfern and Bloswick 1997, Grönqvist *et al.* 2001, Redfern *et al.* 2001). The risk exposure also depends on the person's activity and work task (walking, running, load carrying, lifting, pushing or pulling, etc.), the characteristics of the interacting surfaces (level or inclined surface, surface

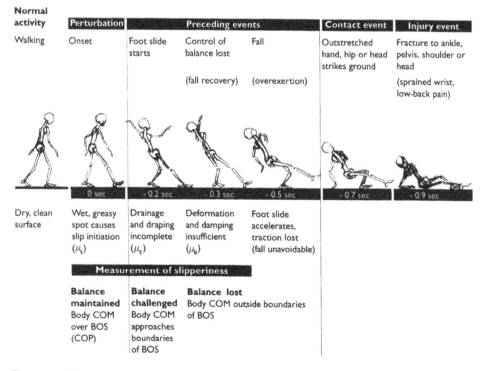

Figure 1. A conceptual scenario of the events leading to slipping and falling after heel contact is presented and the measurement of slipperiness processes prior to and during slipping are shown: static friction coefficient (μ_s), transitional kinetic friction coefficient (μ_t), and steady-state kinetic friction coefficient (μ_k) relate to the shoe/floor interaction, while centre of body mass (COM), base of support (BOS), and centre of foot pressure (COP) relate to postural balance and stability.

R. Grönqvist et al.

discontinuity, floor and footwear materials, etc.), and other adverse conditions that may reduce traction or cause interference (contaminants, obstacles, etc.).

3.3. *Safety criteria*

In addition to mechanical friction measurements using a variety of slip testing devices in the laboratory and in the field (Strandberg 1983, Grönqvist 1999, Chang *et al.* 2001c), human experiments during walking and during other activities have also been performed to estimate shoe/floor and shoe/ice traction and the risk of slipping (Redfern and Rhoades 1996, Grönqvist *et al.* 2001, Redfern *et al.* 2001). Both objective and subjective slip and/or fall predictors, safety criteria, and thresholds have been investigated. The most basic safety criteria are based on two- or three-directional ground reaction force ratios (F_H/F_V in figure 2) and have been named

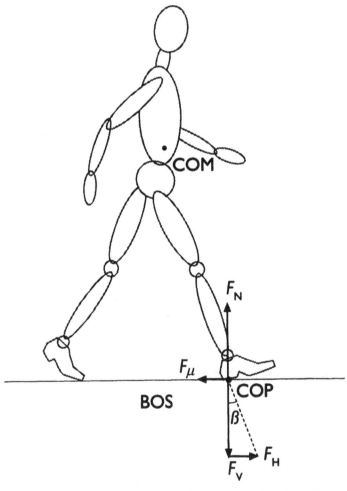

Figure 2. Minimum friction requirement for slip avoidance based on the equilibrium of forces at heel contact: friction force (F_μ), normal force (F_N) and the friction coefficient (μ), as well as the horizontal (F_H) and vertical (F_V) force components applied by the foot are shown together with the locations of the centre of body mass (COM) and the centre of foot pressure (COP).

'friction usage' or 'utilised friction' (Strandberg and Lanshammar 1981, Strandberg *et al*. 1985). Also the terms 'achievable friction' and 'required friction' have been used for the same purpose (McVay and Redfern 1994, Redfern and Rhoades 1996). Required friction is a baseline friction demand for safe gait, typically measured on dry non-slip surfaces (Redfern *et al*. 2001). Utilised friction and its equivalent achievable friction, on the contrary, is independent of relative motion in the shoe/floor interface. Utilised friction is by definition less or equal to required friction.

In order to prevent a slip, the resisting friction force should be at least as high as the horizontal component of the force applied by the foot against the ground. Carlsöö (1962) suggested that walking can be regarded as safe when the measured friction coefficient (i.e. assuming a valid test apparatus is being used) is greater than the ratio of the horizontal and vertical components of the ground reaction force for the actual shoe/floor condition $\mu = \frac{F\mu}{F_N} > \frac{F_H}{F_V}$, (figure 2). In practice, ground reaction force ratios have been calculated either from maximum peak values or from stance time averaged mean values of the horizontal and vertical force components. One of the main concerns with such a definition of safety is that using even the most advanced slip testing devices, it may be difficult to closely reproduce the biomechanics and tribophysics of slipping during gait (Chang *et al*. 2001b, c, Redfern *et al*. 2001). Another concern is that the friction coefficient between surfaces in relative motion is not a constant but a function of normal force and pressure, contact time, sliding velocity, and other parameters (Strandberg 1985).

Several other safety criteria have been proposed in the research literature, for instance, micro-slip and heel slip distance estimates (Perkins 1978, Strandberg and Lanshammar 1981, Leamon and Son 1989, Leamon and Li 1990, Myung *et al*. 1993), subjective slipperiness ratings of footwear and/or floorings (Strandberg *et al*. 1985, Tisserand 1985, Swensen *et al*. 1992, Grönqvist *et al*. 1993, Myung *et al*. 1993), estimates based on heel slip velocities (Strandberg and Lanshammar 1981) or horizontal trunk accelerations (Hirvonen *et al*. 1994), and estimates based on slip/fall probabilities using logistic regression models to relate the observed slip/fall events to the difference between measured and required friction (Hanson *et al*. 1999).

4. Definition of some terms in the conceptual framework

Three distinctive categories of terms in the conceptual framework for the measurement of slipperiness are presented in figure 3. They are related to the causation of slips and falls injuries, to maintaining balance and stability before and after the onset of a foot slide, and to the frictional phenomena in the shoe/floor interface. An attempt is made to define these terms as they appear in the text below.

4.1. *Causation of slips and falls injuries*

The causality of slips and falls injuries is not yet fully understood. Before effective prevention strategies can be put into practice, one must better clarify the accident and injury mechanisms involved (cf. figure 1). The chain or network of events—comprising the exposure to hazards and the initiation of events leading to final injury and disability—need to be identified and analysed in-depth by applying epidemiological, psychological, biomechanical and tribophysical research principles and methodologies.

Wilde (1985) has spoken for a risk homeostasis in the context of traffic safety by addressing the issue of human adaptation seeking to maintain a stable level of perceived risk. More generally, performance is likely to be maintained close to the

Disability is 'any restriction or lack of the ability to perform an activity in the manner or within the range considered normal' (WHO 1980: 143). A person's 'disability status' or 'activity of restriction' describes how the impairment affects activities at work or home or participation in sports. *Handicap* is 'a disadvantage for an impairment or a disabled individual that limits or prevents the fulfilment of normal life roles' (WHO 1980: 183).

4.2. *Gait and balance*

Balance may be externally challenged through the environment and the work task (design factors and conditions affecting traction, etc.), or internally through misperception of and lacking adaptation to external changes in conditions (due to unawareness of the risk or due to risk-taking behaviour, etc.). The biomechanics of slipping is specifically discussed in a paper by Redfern *et al.* (2001). The role of sensory control of posture and balance is discussed in a subsequent paper by Grönqvist *et al.* (2001).

Foot/floor interactions are related to posture and balance during gait as well as many other situations such as manual materials handling tasks that often involve much greater shear forces and frictional demands than locomotion (Grieve 1983). Winter (1995: 3) defined *posture* as 'a term to describe the orientation of any body segment relative to the gravitational vector' and that 'it is an angular measure from the vertical'. Winter (1995: 3) defined *balance* as 'a generic term to describe the dynamics of body posture to prevent falling' and that 'balance is related to the inertial forces acting on the body and the inertial characteristics of body segments'.

Redfern and Schuman (1994) referred to studies where *balance* was defined as 'regulation of dynamic movement of body segments about a supporting joint or base of support'. 'Standing balance' according to Redfern and Schuman is clinically quantified by measuring static postural oscillations (sway), in the antero-posterior and medio-lateral planes, and by recovery of posture on a moving platform. This 'sway' is assumed to reflect our balancing abilities (i.e. the relationships between the body's centre of mass or centre of pressure and a fixed base of support). In standing, the *base of support* (BOS) is defined as 'the area in contact with the supporting surface within the outline of both feet' (Holbein and Chaffin 1997). Pai and Patton (1997) described that the BOS is the roughly trapezoidal contact area between the feet and the floor, in which the resultant ground reaction force is located.

There is no standard definition of 'postural stability' according to Holbein and Chaffin (1997); however, in stable posture the body's centre of mass is within the base of support. Winter (1995: 3) defined the *centre of mass* (COM) as 'a point equivalent of the total body mass in the global three-dimensional spatial reference system with the gravitational vector as the primary reference' and that 'it is the weighed average of the COM of each body segment in that space'. *Centre of pressure* (COP) has been defined as 'the point location of the vertical ground reaction force vector' (Winter 1995: 3). Holbein and Chaffin (1997) found that 'functional stability limits' for persons standing in extreme postures were less than the theoretical maximum indicated by the BOS. Thus, the persons cannot actually reach their BOS limits especially while holding external loads.

Maintaining 'balance during gait' is a much more complex task than during standing, because the BOS is changed from step to step or due to a foot slide, if the surface becomes slippery. Thus balance during gait (and during slipping) includes controlling movements of the body's COM and the placement of the BOS. The position of the body's COM is controlled within the 'limits of stability' during the stance phase (double-leg support) and the swing phase (one-leg support) by continually establishing a new BOS (Redfern and Schuman 1994). The exact size and shape of this 'stability region' largely depends on the mechanical properties of the body and the response latencies of the neuromuscular system for postural control. Ultimately, safe performance is governed by each individual's momentary limits of stability, which may vary over time and place. For attaining 'dynamic stability' of biped locomotion, the step length and the cycle time control appear to provide the most important mechanism (Gubina *et al.* 1974).

4.3. *Frictional phenomena between shoes and surfaces*

The role of friction is crucial in understanding the causation of slips and falls (Grönqvist 1999, Chang *et al.* 2001b). The frictional mechanisms during standing, walking and many other work tasks comprise an interaction between an elastic shoe sole, an interfacial contaminant, and a more rigid floor surface. The friction force is not conservative, its value is governed by the tangential stresses arising in the relative motion of bodies. Friction due to adhesion and/or hysteresis must be present to facilitate safe movement. *Adhesion* is a surface component of friction, due to a molecular-kinetic, thermally-activated dissipative stick-slip mechanism (Moore 1975: 4). The energy parameters characterising a solid and its interface should be considered in friction from a physicochemical standpoint. Surface energy relates to adhesion and depends on the nature and structure of a solid. The *surface energy* is caused by a difference in molecular interactions on the surface of a solid and in the adjacent region compared to the bulk material (Shpenkov 1995: 1). The surface energy is anisotropic (i.e. has directionally different physical properties such as crystals) and has a definite value corresponding to the thermodynamic equilibrium of the surface and its saturated vapour.

Hysteresis is an irreversible and delayed response during an elastic contract stress cycle due to damping in materials that leads to energy dissipation (Chang *et al.* 2001b). Hysteresis results in an asymmetric deformation and pressure distribution during relative sliding at the contact interface (Moore 1975: 5), and it may be termed a bulk phenomenon at the sliding interface. Kummer (1966) showed that both the adhesion and the hysteresis components of rubber friction are manifestations of the same basic viscoelastic energy dissipation mechanism.

Wear and abrasion are caused by local stress and mechanical interlocking between a rubber sole and a rigid floor surface and may cause very high frictional forces (Moore 1975: 12). Wear is the loss of substance as a result of relative motion of the bodies and involves time-dependent and temperature-dependent properties related to their viscoelastic behaviour. The wear mechanisms for rubber-like polymers are: (1) interfacial surface wear (involving high energy densities) such as adhesive wear and tribochemical (corrosion, fretting) wear; and (2) cohesive wear (involving smaller energy densities) such as fatigue wear and abrasive wear (Zum Gahr 1987: 292–293).

The value of the friction force is governed by the tangential stresses arising in the relative motion of bodies. Since the contact surface is wavy and rough (Chang *et al.* 2001a), a contact occurs only on discrete spots which in combination make the *real contact area* (Shpenkov 1995: 25). The *coefficient of friction* (μ) is simply defined as a numerical ratio of the orthogonal forces between the interacting surfaces (F_μ/F_N, i.e. frictional force divided by normal force). Often *slip resistance* (traction, grip) is quantitatively determined by the friction coefficient (Strandberg 1985, Tisserand 1985). An adequate static friction coefficient (μ_s) is assumed to prevent the initiation of a foot slide, for instance during early heel contact as shown in figure 1. A sufficient transitional kinetic friction coefficient (μ_t) and steady-state kinetic friction coefficient (μ_k) are assumed to prevent a still controllable foot slide from proceeding into an uncontrolled slide, a lost balance, and a fall-related or any other injury (Grönqvist 1999, Chang *et al.* 2001b, c).

Strandberg (1985) pointed out that similar frictional mechanisms to the ones valid for a rolling pneumatic tyre on a wet roadway (Moore 1975) also seem to determine traction and grip during gait. These mechanisms comprise (figure 4): (1) the squeeze-film process and drainage capability of the shoe/floor contact surface (lubrication, contamination); (2) the draping and deformation of the shoe heel and sole about the asperities of the underfoot surface (damping, hysteresis); and (3) the true contact and traction between the interacting surfaces (adhesion, wear and abrasion).

5. Conclusions

The measurement of slipperiness may simply comprise an estimation of the slipping hazard exposures that initiate the chain of events ultimately causing an injury. Mechanical slip testing approaches have been readily utilised to measure slipperiness in terms of friction or slip resistance but with conflicting outcomes (Chang *et al.* 2001b, c). The measurement of slipperiness processes may also need to consider the human capacity to anticipate slipperiness and adapt to slippery and unsafe environments for avoiding a loss of balance, which precedes every fall-related injury or any other injury caused by a foot slide. Biomechanical or human-centred measurements may be utilised for such an approach, including an evaluation of relevant safety criteria and a validation of slip test devices (Grönqvist *et al.* 2001, Redfern *et al.*, 2001).

Several underlying causes and mechanisms of slips and falls, some discussed here, must be better understood before the most effective preventive strategies can be put into practice (Courtney *et al.* 2001, Redfern *et al.* 2001). Contact time related

Drainage, draping and true contact in the shoe-floor interface

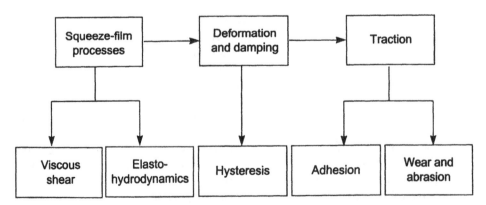

Figure 4.   Basic frictional mechanisms during gait; drainage (squeeze-film processes), draping (deformation and damping) and true contact (traction) in the shoe/floor interface.

variation of ground reaction forces and their relation to frictional surface phenomena during safe gait and during slip/fall incidents need to be determined (Chang *et al.* 2001a, b, c). Muscle responses as well as forces and moments acting on the human body during attempts to recover from slip and fall perturbations need to be evaluated (Redfern *et al.* 2001). For this analysis, the position of the body's centre of mass over the base of support and the point location of the centre of pressure at the feet over the surface of the area in contact with the ground must be calculated.

Biomechanical and tribophysical models for predicting slips and falls are needed. These may be equally beneficial for predicting mechanical stresses on the musculoskeletal system. Optimal safety criteria and safety thresholds for preventing slips and falls cannot be defined without detailed knowledge of the measurement of slipperiness processes. An improved understanding of the measurement of slipperiness paradigm would seem to involve an integration of the methodologies applied in several disciplines, among others, injury epidemiology, psychophysics, biomechanics, motor control, materials science and tribology. Progress has been made in each of these areas during the last several decades; however, much remains to be done. The state-of-the-art of the measurement of slipperiness, the gaps in our knowledge and some future directions for research will be demonstrated in this series of research papers.

Acknowledgements
The authors are grateful to Milja Ahola for her assistance in designing the artwork for figures and to Patti Boelsen for her assistance in preparing the manuscript. We wish to thank the 'opponent' group members (James Collins, Mary Ann Holbein-Jenny, Derek Manning and Gary Sorock) for their review of the preliminary manuscript presented at the Measurement of Slipperiness symposium in Hopkinton, MA (27 – 28 July 2000). The authors also wish to thank all the other reviewers of the manuscript (Patrick Dempsey, Krystyna Gielo-Perczak, Mary Lesch, Simon Matz and Gordon Smith) during the course of this work. This manuscript was completed in part during Dr. Grönqvist's tenure as a researcher at the Liberty Mutual Research Center for Safety and Health.

References
ANDERSSON, R. and LAGERLÖF, E. 1983, Accident data in the new Swedish information system on occupational injuries, *Ergonomics.*, **26**, 33–42.
ANDERSSON, R. and SVANSTRÖM, L. 1980, Olycksfall i arbete – analys och åtgärder (in Swedish). (Stockholm: Almqvist and Wiksell).
BURDORF, A., SOROCK, G., HERRICK, R. F. and COURTNEY, T. K. 1997, Advancing epidemiologic studies of occupational injury – approaches and future directions, *American Journal of Industrial Medicine*, **32**, 180–183.
CAPPOZZO, A. 1991, The mechanics of human walking, in A. E. Patla (ed.), *Adaptability of Human Gait: Implications for the Control of Locomotion* (Amsterdam: Elsevier/North-Holland), 55–97.
CARLSÖÖ, S. 1962, A method for studying walking on different surfaces, *Ergonomics*, **5**, 271–274.
CHANG, W.-R., KIM, I.-J., MANNING, D. and BUNTERNGCHIT, Y. 2001a, The role of surface roughness in the measurement of slipperiness, *Ergonomics*, **44**, 1200–1216.

CHANG, W.-R., GRÖNQVIST, R., LECLERCQ, S., MYUNG, R., MAKKONEN, L., STRANDBERG, L.,
    BRUNGRABER, R., MATTKE, U. and THORPE, S. 2001b, The role of friction in the
    measurement of slipperiness, Part 1: Friction mechanisms and definition of test
    conditions, *Ergonomics*, **44**, 1217–1232.
CHANG, W.-R., GRÖNQVIST, R., LECLERCQ, S., BRUNGRABER, R., MATTKE, U., STRANDBERG, L.,
    THORPE, S., MYUNG, R., MAKKONEN, L. and COURTNEY, T. K. 2001c, The role of friction
    in the measurement of slipperiness, Part 2: Survey of friction measurement devices,
    Ergonomics, **44**, 1233–1261.
CIRIELLO, V. M., McGORRY, R. W. and MARTIN, S. E. 2001, Maximum acceptable horizontal
    and vertical forces of dynamic pushing on high and low coefficient of friction floors,
    International Journal of Industrial Ergonomics, **27**, 1–8.
COURTNEY, T. K., SOROCK, G. S., MANNING, D. P., COLLINS, J. W. and HOLBEIN-JENNY, M. A.
    2001, Occupational slip, trip, and fall-related injuries—can the contribution of
    slipperiness be isolated? *Ergonomics*, **44**, 1118–1137.
ENG, J. J., WINTER, D. A. and PATLA, A. E. 1994, Strategies for recovery from a trip in early
    and late swing during human walking. *Experimental Brain Research*, **102**, 339–349.
GRIEVE, D. W. 1983, Slipping due to manual exertion, *Ergonomics*, **26**, 61–72.
GRÖNQVIST, R. 1999, Slips and falls, in S. Kumar (ed.), *Biomechanics in Ergonomics* (London:
    Taylor & Francis), 351–375.
GRÖNQVIST, R. and ROINE, J. 1993, Serious occupational accidents caused by slipping, in R.
    Nielsen and K. Jorgensen (eds), *Advances in Industrial Ergonomics and Safety V*
    (London: Taylor & Francis), 515–519.
GRÖNQVIST, R., HIRVONEN, M. and TUUSA, A. 1993, Slipperiness of the shoe-floor interface:
    comparison of objective and subjective assessments. *Applied Ergonomics*, **24**, 258–262.
GRÖNQVIST, R., ABEYSEKERA, J., GÅRD, G., HSIANG, S. M., LEAMON, T. B., NEWMAN, D. J.,
    GIELO-PERCZAK, K., LOCKHART, T. E. and PAI, Y.-C. 2001, Human-centred approaches in
    slipperiness measurement, *Ergonomics*, **44**, 1167–1199.
GUBINA, F., HEMAMI, H. and McGEE, R. B. 1974, On the dynamic stability of biped
    locomotion, *IEEE Transactions on Biomedical Engineering*, **21**, 102–108.
HAGBERG, M., CHRISTIANI, D., COURTNEY, T. K., HALPERIN, W., LEAMON, T. and SMITH, T. J.
    1997, Conceptual and definitional issues in occupational injury epidemiology, *American
    Journal of Industrial Medicine*, **32**, 106–115.
HANSON, J. P., REDFERN, M. S. and MAZUMDAR, M. 1999, Predicting slips and falls considering
    required and available friction, *Ergonomics*, **42**, 1619–1633.
HIRVONEN, M., LESKINEN, T., GRÖNQVIST, R. and SAARIO, J. 1994, Detection of near accidents by
    measurement of horizontal acceleration of the trunk, *International Journal of Industrial
    Ergonomics.*, **14**, 307–314.
HOLBEIN, M. A. and CHAFFIN, D. B. 1997, Stability limits in extreme postures: effects of load
    positioning, foot placement, and strength, *Human Factors*, **39**, 456–468.
HOOGENDOORN, W. E., VAN POPPEL, M. N. M., BONGERS, P. M., KOES, B. W. and BOUTER, L. M.
    1999, Physical load during work and leisure time as risk factors for back pain,
    Scandinavian Journal of Work Environment and Health, **25**, 387–403.
HOOZMANS, M. J. M., VAN DER BEEK, A. J., FRINGS-DRESDEN, M. H. W., VAN DUK, F. J. H. and
    VAN DER WOUDE, L. H. V. 1998, Pushing and pulling in relation to musculoskeletal
    disorders: a review of risk factors, *Ergonomics*, **41**, 757–781.
HSIAO, E. T. and ROBINOVITCH, S. N. 1998, Common protective movements govern unexpected
    falls from standing height, *Journal of Biomechanics*, **31**, 1–9.
KUMMER, H. W. 1966, Unified theory of rubber and tire friction, *Engineering Research Bulletin
    B-94* (Pennsylvania: The Pennsylvania State University).
LAST, J. M. (ed.) 1995, *A Dictionary of Epidemiology*, 3rd edn, International Epidemiological
    Association (New York: Oxford University Press).
LAVENDER, S. A., SOMMERICH, C. M., SUDHAKER, L. R. and MARRAS, W. S. 1988, Trunk muscle
    loading in non-sagittally symmetric postures as a result of sudden unexpected loading
    conditions, *Proceedings of the Human Factors Society 32nd Annual Meeting* (Santa
    Monica, CA: The Human Factors Society), vol. 1, 665–669.
LEAMON, T. B. and LI, K.-W. 1990, Microslip length and the perception of slipping, Paper
    presented at the 23rd International Congress on Occupational Health, 22–28
    September, Montreal, Canada.

LEAMON, T. B. and MURPHY, P. L. 1995, Occupational slips and falls: more than a trivial problem, *Ergonomics*, **38**, 487–498.

LEAMON, T. B. and SON, D. H. 1989, The natural history of a microslip, in A. Mital (ed.), *Advances in Industrial Ergonomics and Safety I* (London: Taylor & Francis), 633–638.

LECLERCQ, S. 1999, The prevention of slipping accidents: a review and discussion of work related to the methodology of measuring slip resistance, *Safety Science*, **31**, 95–125.

LLEWELLYN, M. G. A. and NEVOLA, V. R. 1992, Strategies for walking on low-friction surfaces, in W. A. Lotens and G. Havenith (eds), *Proceedings of the Fifth International Conference on Environmental Ergonomics*, Maastricht, The Netherlands, 156–157.

MANNING, D. P. and SHANNON, H. S. 1981, Slipping accidents causing low-back pain in a gearbox factory, *Spine*, **6**, 70–72.

MANNING, D. P., MITCHELL, R. G. and BLANCHFIELD, L. P. 1984, Body movements and events contributing to accidental and nonaccidental back injuries, *Spine*, **9**, 734–739.

MANNING, D. P., AYERS, I., JONES, C., BRUCE, M. and COHEN, K. 1988, The incidence of underfoot accidents during 1985 in a working population of 10,000 Merseyside people, *Journal of Occupational Accidents*, **10**, 121–130.

McVAY, E. J. and REDFERN, M. S. 1994, Rampway safety: foot forces as a function of rampway angle, *American Industrial Hygiene Association Journal*, **55**, 626–634.

MOORE, D. F. 1975, *The friction of pneumatic tyres* (Amsterdam: Elsevier).

MURPHY, P. L. and COURTNEY, T. K. 2000, Low back pain disability: relative costs by antecedent and industry group, *American Journal of Industrial Medicine*, **37**, 558–571.

MYUNG, R., SMITH, J. and LEAMON, T. B. 1993, Subjective assessment of floor slipperiness, *International Journal of Industrial Ergonomics*, **11**, 313–319.

NASHNER, L. M. 1980, Balance adjustments of humans perturbed while walking, *Journal of Neurophysiology*, **44**, 650–664.

PAI, Y.-C. and PATTON, J. 1997, Center of mass velocity-position predictions for balance control, *Journal of Biomechanics*, **30**, 347–354.

PERKINS, P. J. 1978, Measurement of slip between the shoe and ground during walking, in C. Anderson and J. Sene (eds), *Walkway Surfaces: Measurement of Slip Resistance*, ASTM STP 649 (Baltimore, MD: American Society for Testing and Materials), 71–87.

RASMUSSEN, J. 1997, Risk management in a dynamic society: a modelling problem, *Safety Science*, **27**, 183–213.

REDFERN, M. S. and BLOSWICK, D. 1997, Slips, trips, and falls, in M. Nordin, G. Andersson and M. Pope (eds), *Musculoskeletal Disorders in the Workplace* (Location: Masby-Year Inc.), 152–166.

REDFERN, M. S. and RHOADES, T. P. 1996, Fall prevention in industry using slip resistance testing, in A. Bhattacharya and J. D. McGlothlin (eds), *Occupational Ergonomics, Theory and Applications* (New York: Marcel Dekker), 463–476.

REDFERN, M. S. and SCHUMAN, T. 1994, A model of foot placement during gait, *Journal of Biomechanics*, **27**, 1339–1346.

REDFERN, M. S., CHAM, R., GIELO-PERCZAK, K., GRÖNQVIST, R., HIRVONEN, M., LANSHMMAR, H., MARPET, M., PAI, Y.-C. and POWERS, C. 2001, Biomechanics of slips, *Ergonomics*, **44**, 1138.–1166.

ROUHIAINEN, V. 1990, *The quality assessment of safety analysis*, Publication 61 (Espoo: Technical Research Centre of Finland)

SCHÖN, G. 1980, What is meant by risk? Basic technical views for the initiation and applications of safety legislation, *Journal of Occupational Accident*, **2**, 273–281.

SHPENKOV, G. P. 1995, Friction surface phenomena, in D. Dowson (ed.), *Tribology Series 29* (Amsterdam: Elsevier).

STOBBE, T. J. and PLUMMER, R. W. 1988, Sudden-movement/unexpected loading as a factor in back injuries, in F. Aghazadeh (ed.), *Trends in Ergonomics/Human Factors V* (Amsterdam: Elsevier/North-Holland), 713–720.

STRANDBERG, L. 1983, Ergonomics applied to slipping accidents, in T. O. Kvålseth (ed.), *Ergonomics of Workstation Design* (London: Butterworths), 201–228.

STRANDBERG, L. 1985, The effect of conditions underfoot on falling and overexertion accidents, *Ergonomics*, **28**, 131–147.

STRANDBERG, L. and LANSHAMMAR, H. 1981, The dynamics of slipping accidents, *Journal of Occupational Accidents*, **3**, 153–162.

16 R. *Grönqvist* et al.

STRANDBERG, L., HILDESKOG, L. and OTTOSON, A.-L. 1985, Footwear friction assessed by walking experiments, *VTIrapport 300 A* (Linköping: Väg- och trafikinstitutet).

SWENSEN, E., PURSWELL, J., SCHLEGEL, R. and STANEVICH, R. 1992, Coefficient of friction and subjective assessment of slippery work surfaces, *Human Factors*, **34**, 67–77.

TAYLOR, D. H. 1987, The role of human action in man machine systems, in J. Rasmussen, K. Duncan and J. Leplat (eds), *New Technology and Human Error* (New York: Wiley).

TISSERAND, M. 1985, Progress in the prevention of falls caused by slipping, *Ergonomics*, **28**, 1027–1042.

WALLER, J. A. 1985, *Injury Control. A guide to the Causes and the Prevention of Trauma* (Massachusetts/Toronto: Lexington Books).

WILDE, G. J. S. 1985, Assumptions necessary and unnecessary to risk homeostasis, *Ergonomics*, **28**, 1531–1538.

WINTER, D. A. 1991, *The Biomechanics and Motor Control of Human Gait: Normal, Elderly, and Pathological*, 2nd edn (Ontario, Canada: University of Waterloo).

WINTER, D. A. 1995, *ABC: Anatomy, Biomechanics and Control of Balance during Standing and Walking* (Ontario, Canada: University of Waterloo).

WORLD HEALTH ORGANISATION 1980, International classification of impairments, disabilities, and handicaps (Geneva: World Health Organisation).

ZUM GAHR, K. H. 1987, Microstructure and wear of materials, *Tribology Series 10* (Amsterdam: Elsevier).

CHAPTER 2

Occupational slip, trip, and fall-related injuries—can the contribution of slipperiness be isolated?

THEODORE K. COURTNEY†*, GARY S. SOROCK†, DEREK P. MANNING‡, JAMES W. COLLINS§ and MARY ANN HOLBEIN-JENNY¶

†Liberty Mutual Research Center for Safety and Health, 71 Frankland Road, Hopkinton, MA 01748, USA

‡341 Liverpool Road, Birkdale, Southport, Merseyside PR8 3DE, UK

§Division of Safety Research, National Institute for Occupational Safety and Health, Morgantown, WV 26505, USA

¶Graduate School of Physical Therapy, Slippery Rock University, Slippery Rock, PA 16057, USA

Keywords: Falls; Occupational injuries; Slipping; Epidemiology; Surveillance.

To determine if the contribution of slipperiness to occupational slip, trip and fall (STF)-related injuries could be isolated from injury surveillance systems in the USA, the UK and Sweden, six governmental systems and one industrial system were consulted. The systems varied in their capture approaches and the degree of documentation of exposure to slipping. The burden of STF-related occupational injury ranged from 20 to 40% of disabling occupational injuries in the developed countries studied. The annual direct cost of fall-related occupational injuries in the USA alone was estimated to be approximately US$6 billion. Slipperiness or slipping were found to contribute to between 40 and 50% of fall-related injuries. Slipperiness was more often a factor in same level falls than in falls to lower levels. The evaluation of the burden of slipperiness was hampered by design limitations in many of the data systems utilised. The resolution of large-scale injury registries should be improved by collecting more detailed incident sequence information to better define the full scope and contribution of slipperiness to occupational STF-related injuries. Such improvements would facilitate the allocation of prevention resources towards reduction of first-event risk factors such as slipping.

1. Introduction

Slip, trip, and fall (STF)-related morbidity and mortality are considerable in many developed countries. In the USA, for example, slips and falls are the second largest source of unintentional injury mortality each year (Fingerhut *et al.* 1998). Overall in 1997 in the USA falls were the leading external cause of medically-attended, non-fatal unintentional injuries with 11.3 million episodes reported at an age-adjusted rate of 43.1 per 1000 persons (Warner *et al.* 2000). Slips and falls were also the leading reason for unintentional injury emergency department visits comprising 21% of such visits (National Safety Council 1998).

---

*Author for correspondence. e-mail: theodore.courtney@libertymutual.com

Occupationally, Courtney and Webster (1999) recently reported that in the USA slips and falls are associated with the most severely disabling, sudden-onset occupational injuries including fractures. Leamon and Murphy (1995) estimated that annual, direct, per capita costs of occupational injuries due to slips and falls ranged from US$50 to US$400 per worker depending on the industry group considered. Substantial losses have also been reported in the UK, Finland, and Sweden (Manning 1983, Manning *et al.* 1988, Grönqvist and Roine 1993, National Safety Council 1995, Kemmlert and Lundholm 1998).

It is commonly assumed that slipperiness and slipping are major contributors to the STF injury burden. Across national boundaries, the registries that record injury data often differ substantially in their characteristics, focus, and *raison d'être* making international comparisons difficult (Hagberg *et al.* 1997). However, while the limitations of registry data may not permit direct combination, elements from various sources can be compared to attempt to improve the overall understanding of a workplace hazard such as slipperiness.

The primary purpose of the present paper was to determine if, and to what extent, the contribution of slipperiness and slipping could be isolated in various injury surveillance systems in the USA, the UK and Sweden. To provide the necessary foundation for this analysis, the characteristics of each system are presented along with concise analyses of the frequency and severity of the STF-related occupational injuries captured.

2.   An overview of the data sources

Seven injury registries were selected to provide a variety of perspectives for the countries chosen as well as approaches to data collection and reporting. Interest lay in observing the STF problem from several different vantage points. It was believed that this approach would afford opportunities to discern the contribution of slipperiness to the STF burden.

The data are presented in four sections. Section 2 introduces the designs and characteristics of various surveillance systems. Section 3 summarises the STF injuries in each of the surveillance systems. Section 4 presents the available data on nature of injury with a specific focus on the most disabling injuries from systems offering fatal and non-fatal perspectives on disability (or cost as a surrogate for disability). Finally, section 5 examines the available data related to slipperiness from the systems that supported such an analysis.

2.1. *United States Census of Fatal Occupational Injuries (CFOI)*

The United States Bureau of Labour Statistics (BLS) administers the Census of Fatal Occupational Injuries (CFOI). The CFOI collects information about all fatal work injuries in the USA including those incurred by wage and salary workers in private industry and government, and self-employed workers. This surveillance system has collected occupational fatality data nationwide since 1992 and is considered to be a fairly comprehensive census because it gathers information from multiple sources including death certificates, coroner, medical examiner, and autopsy reports, workers' compensation reports, Occupational Safety and Health Administration (OSHA) reports, motor vehicle crash reports, employer questionnaires, and newspaper articles.

Owing to co-operative agreement restrictions between New York City and the BLS, data from New York City were not included in the analyses of fatal occupational falls in this report. This explains small differences between totals reported here and totals reported in BLS publications.

Descriptive data on occupational STF-related fatalities were available in the CFOI. The BLS Occupational Injury and Illness Classification Scheme also known as OIICS (USDOL-BLS 1992) is used, and falls are categorised into three major groups. Falls on the same level occur when the point of contact with the source of injury is on the same level or above the surface supporting the injured person. Falls to a lower level occur when the point of contact with the source of injury is below the level of the surface supporting the injured person. Finally, jumps to a lower level occur when the injured person leaps from an elevation voluntarily, even if the jump is to avoid an uncontrolled fall or other injury.

2.2. *US Survey of Occupational Injuries and Illness (SOII)*

The BLS also conducts an annual Survey of Occupational Injuries and Illnesses (SOII). From a sample of roughly 200 000 establishments in private industry, the BLS collects employer injury and illness reports and estimates the overall US occupational injury and illness experience (Murphy *et al.* 1996, USDOL-BLS 1997). Agricultural establishments with fewer than 11 employees, self-employed persons, and federal government employees are excluded from the survey. Workers in state and local governments are not included in national estimates.

Annual data on non-fatal injuries associated with slips, trips and falls became available in 1992 when the BLS introduced an expanded SOII method that collected additional detailed data on non-fatal cases with 1 or more days-away-from-work (DAW) (Courtney and Webster 1999). Like the CFOI, the SOII uses the BLS' OIICS coding approach. Data are available on injuries arising from falls to a lower level, falls on the same level, jumps to lower level, and slips, trips or loss of balance without a fall.

2.3. *US National Electronic Injury Surveillance System (NEISS)*

The National Electronic Injury Surveillance System (NEISS) database is a collaborative project between the National Institute for Occupational Safety and Health (NIOSH) and the Consumer Product Safety Commission (CPSC). Initially designed to collect information on injuries from consumer products, it also collects data for all occupational injuries regardless of the source of injury. NEISS comprises 91 hospitals selected as a stratified probability sample of all hospitals in the USA. Any injury that is determined to be work-related during review of the emergency department (ED) records is included in the data. Each injury case in the sample is assigned a statistical weight based on the hospital's probability of selection and used in extrapolation techniques to calculate national estimates (US Centers for Disease Control and Prevention 1998).

2.4. *Data from a large workers' compensation provider in the USA*

One workers' compensation provider (WCP) in the USA covers approximately 10% of the private insurance market with a wide distribution of coverage nationally. Analyses of WCP claims data have been published previously including Leamon and Murphy (1995), Dempsey and Hashemi (1999), and Murphy and Courtney (2000), which addressed various aspects of STF morbidity. The population of claims includes claims from each state of the USA and the District of Columbia, except the six states with monopolistic, state fund workers' compensation systems: Nevada,

North Dakota, Ohio, Washington, West Virginia, and Wyoming. Claims data for 1996 were retrieved in August 1999, allowing a minimum 2.5 year window for claim cost development. Claims involving falls from height and falls on the same level were identified using antecedent event categories based on a proprietary 'cause' code assigned by the WCP claims department. The characteristics of WCP data are further described in Murphy *et al.* (1996).

2.5. *US National Health Interview Survey (NHIS)*

The National Health Interview Survey (NHIS) is the principal source of information on the health of the civilian population of the USA and is one of the major data collection programs of the National Center for Health Statistics (NCHS). Since 1960, NCHS has utilised the NHIS to monitor trends in illness and disability and to track progress toward achieving national health objectives. NHIS data are collected annually from approximately 43 000 households including about 106 000 persons. The annual response rate of NHIS is greater than 90% of the eligible households in the sample.

In 1997 a substantially revised NHIS was fielded. This revision greatly improved the ability of the NHIS to provide important health information especially for injuries. A major strength of this survey lies in the ability to cross-tabulate occupational injury data by many demographic and socio-economic characteristics. The NHIS is also population-based, and thus can provide data on all persons injured at work regardless of workers' compensation coverage, industry or employment status. The NHIS utilises the *International Classification of Diseases, 9th Revision, Clinical Modification* for coding injury causes and health outcomes (US Public Health Service and Health Care Financing Administration 1991).

The NHIS covers the civilian non-institutionalised population of the USA living at the time of the interview. However, there are several segments of the population which are not included in the sample or in the estimates from the survey. Exclusions are: patients in long-term care facilities, persons on active duty with the Armed Forces (although their dependents are included), and US nationals living in other countries.

2.6. *UK Health and Safety Statistics*

Employers in the UK are required to report any accidents on their premises that cause absence from work for more than 3 days to the national Health and Safety Executive (HSE). There are three categories of severity that are mutually exclusive: fatal, major, and over 3-day accidents (absence from work for more than 3 days). Major injuries include all fractures and dislocations except those involving fingers, thumbs and toes. They also include hospital admissions of more than 24 h.

Annual statistics issued by HSE provide accident numbers and rates per 100 000 employed (excluding the self-employed) for five main occupational groups (HSE 1999). There are 17 'kinds' of accident including 'slips, trips and falls' which are subdivided into slips, trips and falls on the same level, falls from a height of up to 2 m and falls from over 2 m. All fatal accidents are reported. Accidents involving employees travelling to and from work are not collected in the UK.

2.7. Swedish Information System Occupational Injuries and Diseases (ISA)

The *information system om arbetsskador* (ISA) or information system on occupational injuries and diseases was instituted in 1979 under the authority of the National

Board of Occupational Safety and Health (Andersson and Lagerlof 1983). ISA collects data on occupational accidents and work-related diseases reported under the Work Injury Insurance Act. In Sweden, all economically active persons are insured for occupational injuries. A case is registered in ISA only if the injured person is absent from work for at least 1 day after the day of the accident. Commuting accidents and self-employed and military populations are included. It is computer-based and registers accidents with a sequential description model allowing several events to be recorded for each accident including the injury event, the contact event and the preceding events (Swedish Work Environment Authority and Statistics Sweden 2000).

3. An examination of the scope of the problem

3.1. *US Census of Fatal Occupational Injuries, 1992–1998*

Over the 7-year period from 1992 to 1998, there were 4507 work-related fatal falls in the USA (excluding New York City). The number of fatal occupational falls steadily increased in the USA from 584 in 1992 to 683 in 1998. Of the 4507 work-related fatal falls that occurred during this period, 3946 (88%) occurred when the worker fell to a lower level, 364 (8%) occurred to workers who fell and struck the surface that was supporting them, 43 (1%) of the workers jumped to their death, and the nature of the fall was not specified in 121 (3%) of the deaths.

The male to female ratio varied by type of fall ranging from about 30 to 1 for falls from ladders and scaffolds to about 1 to 1 for falls on the same level. Figure 1 presents the mortality rates for all occupational falls by age of victim. While workers from 25 to 54 years of age accounted for 65% of fall-related deaths, figure 1 shows that occupational fall mortality rates increased substantially with age peaking in workers 65 years of age and older.

With regard to the 364 fatal 'falls on the same level', 255 (70%) of the fatalities occurred when the worker fell to the floor, walkway, or other surface, 74 (20%) when

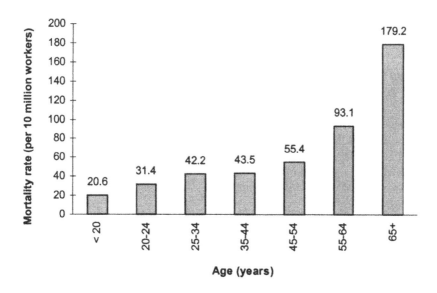

Figure 1.   US occupational fall mortality rate by age 1992–1998 (*n* = 4 507), CFOI.

the worker fell onto or against objects, and in 35 (10%) of the cases the surface or object the worker fell onto was not specified. Common scenarios involved a worker slipping, tripping, or falling and striking their head on a floor, parking lot, concrete, or other rigid surface with sufficient force to cause a fatal brain injury. The incidents where the decedent fell against an object were commonly described as a worker falling against a sharp or pointed object that impaled the worker.

3.2. *US Survey of Occupational Injury and Illness, 1996*

In 1996, the BLS estimated that there were 330 913 non-fatal, fall-related injuries involving 1 or more days away from work (DAW). A total of 30% of these were attributed to falls to a lower level with 66% to falls on the same level, and 4% to jumps to a lower level. The BLS data also indicated that there were an additional 59 328 injuries related to slipping, tripping or loss of balance *without* a resulting fall. These latter injuries comprised 30% of the bodily reaction event category under which they were classified. (Bodily reaction includes slips and trips without falls, but also other injuries due to bodily motion or reaction not involving external objects such as climbing, twisting, running, crawling, etc.). Combining these two estimates provided an estimate of the total burden of disabling occupational STF-related injuries in the USA. Overall, there were an estimated 390 241 disabling STF-related injuries, which accounted for 21% of all disabling injuries reported to the BLS in 1996. STF-related injuries also accounted for 48% of disabling sprains and strains and for 46% of disabling fractures.

Table 1 presents the percentage distribution of fall-related DAW occupational injuries by selected characteristics for 1996. The non-falling injuries noted above were excluded since details were only available at the aggregated bodily reaction category level. Here it may be discerned that while men and women were roughly equivalent in numbers injured in same level falls, men were more likely to be involved in falls to a lower level. Compared to all disabling injuries, disabling fall victims were more likely to be over the age of 55 years. Among industries the Services sector accounted for the largest proportion of STF-related injuries. Interestingly, both falls on the same level and falls from a height had higher median DAW (7 and 10 days, respectively) than the median for all DAW cases (5 days).

3.3. *US National Electronic Injury Surveillance System, 1998*

Data in table 2 are National Electronic Injury Surveillance System (NEISS)-weighted estimates of the number of fall-related occupational injuries treated in US emergency departments in 1998 and were obtained from the National Institute for Occupational Safety and Health, Division of Safety Research, in Morgantown, WV (Jackson and Hixon 2000). The BLS occupational coding scheme (OIICS) was used here as well. The unit of analysis in NEISS was the injury episode not the individual; hence, if someone had two unrelated falls collected in the sample, each incident was counted in table 2.

In 1998 the NEISS-weighted estimates indicated that 550 592 occupational injuries arising from falls were treated in US emergency departments. An additional 49 661 injuries resulted from slips, trips and losses of balance without a fall.

The data suggest a ratio of 2:1 for falls on the same level compared to falls to a lower level. This sample consists of fall events serious enough to warrant a visit to a hospital emergency department. For falls to a lower level and falls on the same level, the trunk, including the shoulder and groin, was injured about as often as the lower

Table 1. Percentage distribution of non-fatal, occupational falls involving days away from work by selected characteristics in the USA, BLS SOII 1996.

Characteristic	All events n = 1 880 525	All falls* n = 330 913	Falls to lower level n = 98 544	Falls on same level n = 219 416
Gender				
Male	65.9	59.7	78.8	49.7
Female	33.0	39.8	20.6	49.8
Age (years)				
14 – 15	0.1	0.3	0.2	0.3
16 – 19	3.9	3.3	2.5	3.7
20 – 24	12.3	9.4	10.6	8.5
25 – 34	29.8	25.3	26.8	24.2
35 – 44	27.2	26.3	27.8	25.8
45 – 54	16.2	20.1	18.1	21.3
55 – 65	6.7	10.8	9.4	11.7
65 or more	0.9	2.1	1.5	2.4
Industry				
Agriculture, forestry, fishing	2.0	2.1	2.9	1.6
Mining	0.8	0.8	1.5	0.5
Construction	9.7	11.5	21.7	6.6
Manufacturing	24.6	16.6	14.8	17.1
Transport, public utilities	11.9	13.0	17.3	11.0
Wholesale trade	7.7	6.3	8.0	5.3
Retail trade	17.1	20.0	11.9	23.8
Finance, insurance, real estate	2.3	3.6	3.7	3.7
Services	23.9	26.0	18.1	30.4
Number of days away				
1	16.7	13.9	11.7	15.0
2	13.1	11.9	11.5	12.1
3 – 5	20.6	18.7	16.9	19.6
6 – 10	13.2	13.1	12.2	13.4
11 – 20	11.7	21.3	12.8	12.1
21 – 30	6.2	6.8	7.0	6.9
31 or more	18.5	23.4	28.0	21.1
Median days away from work	5	7	10	7

* Includes jumps to lower level (approximately 4%).

Table 2. Weighted estimates from NEISS for US emergency department-treated, fall-related, occupational injury events, 1998.

Slip/trip/fall events	Number of injuries	95% confidence interval	Percentage of all STF
Fall on same level	314 027	± 56 315	57
Fall to lower level	164 519	± 29 602	30
Slip/trip without a fall	49 661	± 11 781	9
Jump to a lower level	8 410	± 2 901	2
Unspecified or not elsewhere classified fall	13 975	± 3 641	2
Total	550 592	± 100 066	100

extremities. For slips, trips and losses of balance without a fall, the lower extremities were injured twice as often as the trunk.

3.4. *WCP fall-related claims, 1993–1998*

There were 192 409 claims for injuries due to falls to a lower level from 1993 to 1998. There were an additional 441 745 claims for injuries related to falls on the same level. Same level falls accounted for 70% of the falls claim experience with falls to the lower level comprising 30%. Overall, lower level and same level falls accounted for 5.2% and 12.0% of all workers' compensation claims, respectively, for a combined contribution of 17.2% of all claims filed during the period.

The 634 154 claims filed from 1993 to 1998 accounted for a combined cost of more than US$3.4 billion (25% of all workers' compensation costs). This figure represents the experience of the largest US workers' compensation insurer. From this figure, the direct financial burden of fall-related occupational injuries for the USA over the period was conservatively estimated based on the method of Webster and Snook (1990, 1994) adjusting the cost experience by the company's market share. The total estimate was more than US$37 billion for the entire period or an average of more than US$6 billion annually.

3.5. *US National Health Interview Survey, 1997/1998*

The NHIS questionnaire was administered in person in a house-to-house survey and directly entered into a laptop computer. This method allowed branching of interview questions depending on interviewee responses. The following screening question was used: 'During the past three months, that is 91 days before today's date, were you or anyone in the family injured seriously enough that you/they got medical advice or treatment?'

In 1997 and 1998, the first 2 years of the revised survey, there were an estimated 1 286 180 fall-related injuries at work. A total of 40% of interviewees reporting a fall-related injury reported falling on the same level while 49% reported falling to a lower level, and 11% reported their fall as occurring in some other manner.

3.6. *UK Health and Safety Statistics—STF in Britain excluding Northern Ireland, 1997/98*

Data for STF injuries in mainland Britain for 1997/98 were extracted from the publication *Health and Safety Statistics 1998/9* (HSE 1999). Table 3 presents the calculated incidence of STF accidents by fall type.

Other data reported by the HSE include the number of injuries in each of five main industrial groups: Agriculture, hunting, forestry and fishing; Extractive and utility supply industries; Manufacturing industries; Construction; Service industries. Table 4 shows the numbers of reported accidents and rates per 100 000 employed (excluding self-employed) in the five main industrial groups. The service and manufacturing industries accounted for the highest numbers of major injuries but the highest incidence rates occurred in extractive industries and construction. The HSE statistics do not delineate the numbers of injuries caused by slipping or by tripping because slips, trips and falls are all collected as a single group.

STF accidents were by far the most numerous kind of accident causing major injuries and they were second in rank order of kind of accident causing over 3 days' absence in all five industrial groups; only 'injuries whilst handling,

Table 3. The incidence of STF accidents in mainland Britain during 1997/1998, HSE.

	Fatal	Major*	Absent over 3 days	Total
STF on the same level	0	8 671	25 883	34 554
STF from height	64	5 382	8 452	13 898
All kinds of accidents reported to HSE	212	29 187	134 789	164 188
Percentage of all reported accidents				
STF on level	–	29.8%	19.2%	
STF from height	30.2%	18.4%	6.3%	

* Major injuries include all fractures and dislocations except those involving fingers, thumbs and toes. Hospital admissions of more than 24 h are also included.

Table 4. Reported accidents and rates (per 100 000 employed) in the main industrial sectors, HSE.

Industry	Agriculture	Extractive	Manufacturing	Construction	Service
Employment*	300 534	215 035	4 058 866	1 009 598	17 291 483
STF, same level					
Fatal	0	0	0	0	0
Major	118 (39.3)	154 (71.6)	2 008 (49.5)	727 (72.0)	5 664 (32.8)
Absent over 3 days	202 (67.2)	731 (339.4)	6 795 (167.4)	1 615 (160.0)	16 540 (95.7)
STF, from height					
Fatal	4 (1.3)	2 (0.9)	13 (0.3)	29 (2.9)	13 (0.7)
Major	154 (51.2)	108 (50.2)	1 238 (30.5)	1 427 (141.3)	2 455 (14.2)
Absent over 3 days	130 (43.3)	214 (99.5)	2 239 (55.2)	1 213 (120.1)	4 656 (26.9)

*Number employed in each sector. Total employment, excluding the self-employed, was 22 875 516.

lifting or carrying' were a more frequent kind of accident causing absence of over 3 days.

3.7. *Swedish Information System Occupational Injuries and Diseases, 1998*
Sweden's ISA recorded 37 914 occupational accidents resulting in a victim missing at least 1 day of work in 1998. Included in this figure are injuries to the gainfully employed including those in military service as well as trainees working for no pay. Detailed information was available from the official statistics (Swedish Work Environment Authority and Statistics Sweden 2000) on 35 848 of these cases, which involved employees and self-employed persons. Of these detailed cases, falls accounted for 7 810 accidents (22% of all the accidents) and led all other types as the most common type of occupational accident. Falls to lower levels accounted for 2 609 falls (33% of fall accidents) while falls on the same level contributed 5 201 (67%).

4. Injury severity and disability
Interest next lay in examining severe STF-related injuries to the extent feasible in the available data. Attention focused on fatal injuries and typical, severe, non-fatal injuries (those with the most days away from work and/or highest claims cost).

4.1. *US Census of Fatal Occupational Injury, 1992–1998*

One hundred and ninety-three (53%) of the fatal injuries sustained from 'falls on the same level' were described as blunt trauma to the head that led to fatal brain injuries. Ninety-one (25%) were intracranial injuries, 79 (22%) were multiple intracranial injuries, and 23 (6%) were cerebral haemorrhages. The next highest proportion of fatal injuries were fractures (65, 17.9%), injuries to internal organs (22, 6.0%), and multiple traumatic injuries (15, 4.1%). A total of 32% ($n = 116$) of the decedents who fell on the same level died on the day of the incident, 232 (64%) survived for 7 days or less, and 33 (9%) of the decedents survived for 98 days or more.

4.2. *US Survey of Occupational Injury and Illness, 1996*

Table 5 describes the 10 most disabling injuries resulting from falls to a lower level and falls on the same level. To enter this table an injury had to have a minimum of at least 500 estimated DAW cases during 1996 (Courtney and Webster, 2001).

4.3. *WCP fall-related severe claims costs, 1993–1998*

For a companion viewpoint to the BLS data on the injuries with the highest median DAW, WCP claims for 1996 were cross-tabulated by body part and nature of injury and then sorted by average expense. Table 6 describes the 10 most costly injuries resulting from falls to a lower level and falls on the same level, respectively. To enter this table an injury type had to have accounted for at least US$500 000 in aggregate costs and have resulted in at least 50 claims filed in 1996. Each injury is ranked

Table 5.   Most disabling occupational injuries due to falls to lower level and on the same level in the USA, BLS SOII 1996.

Fall type	Part of body	Nature of injury	Median DAW	Number of cases
To lower level				
	Pelvic region	Fractures	88	762
	Multiple body parts	Fractures and other injuries	63	1 641
	Multiple body parts	Fractures	51	1 046
	Back	Traumatic injuries, unspecified	40	587
	Leg(s)	Traumatic injuries, unspecified	35	551
	Ankle(s)	Fractures	34	3 047
	Shoulder	Nonspecified injuries and disorders	34	593
	Foot (feet), except toe	Fractures	33	2 552
	Leg(s)	Fractures	31	2 103
	Wrist(s)	Fractures	30	3 361
On same level				
	Ankle(s)	Fractures	35	4 056
	Pelvic region	Fractures	32	1 593
	Multiple trunk location	Nonspecified injuries and disorders	32	1 005
	Shoulder	Fractures	32	945
	Shoulder	Traumatic injuries, unspecified	32	694
	Leg(s)	Fractures	29	4 071
	Back	Dislocations	28	919
	Wrist(s)	Fractures	21	5 534
	Multiple body parts	Fractures	21	560
	Hand(s), except finger	Nonspecified injuries and disorders	20	849

according to its ratio against all other claims in its category (to a lower level or on the same level).

In both instances, the injury types presented did not occur with high frequency but did result in substantial disability. If the inclusion criteria based on frequency and total cost used in this analysis were relaxed, less frequently occurring but even more severe injuries would have been presented, in particular those with trauma to the head and spinal cord. These case types would have been similar to the findings for fatal injuries for falls on the same level presented in section 4.1.

5. Contribution of slipperiness

In this section, interest lay in determining if it was possible to isolate or partially isolate the contribution of slipperiness or slipping to the overall burden of STF injuries. This proved to be difficult as slipperiness itself is a difficult concept to track in such large scale systems. However, at least three of the systems provided data in either narrative or coded form relevant to this question.

5.1. *US Census of Fatal Occupational Injury, 1992–1998*

Slipperiness is not tracked in this data system in summary fashion, so it was necessary to examine narratives. To make the examination more manageable, only

Table 6. Most costly occupational injuries due to falls to lower level and on the same level in the USA, WCP 1996.

Fall type	Part of body	Nature of injury	Cost ratio*	Number of cases
To lower level				
	Multiple body parts	Fracture	10.8	762
	Lower leg	Fracture	5.3	110
	Upper arm	Dislocation	3.1	85
	Wrist	Fracture	2.7	287
	Knee	Fracture	2.7	67
	Ankle	Fracture	2.6	366
	Lower arm	Fracture	2.6	122
	Upper arm	Fracture	2.4	83
	Multiple body parts	All other injuries	2.4	1 322
	Foot	Fracture	2.0	316
On same level				
	Disc	Rupture	13.3	99
	Hip	Fracture	5.3	76
	Lower leg	Fracture	4.1	142
	Knee	Fracture	2.7	159
	Knee	Dislocations	2.4	113
	Upper arm	Fracture	1.8	151
	Ankle	Fracture	1.6	591
	Elbow	Fracture	1.6	252
	Hip	Sprain	1.5	66
	Upper arm	Dislocations	1.3	174

*Cost ratio is the ratio of the average cost of the particular injury to the average cost of all injuries for that particular class of falls (e.g. a cost ratio of 5.3 for lower leg fractures means that for claims involving falls to a lower level, those resulting in lower leg fractures cost on average 5.3 times more than the average for all falls to a lower level.

records involving falls on the same level were selected for detailed review. This category was expected to have a higher proportion of fall-related deaths involving slipperiness.

Narrative descriptions of fatal occupational falls on the same level did not provide antecedent information for 169 (approximately 46%) of the incidents. However, for the remaining 195 (54%), a manual review of the narrative descriptions identified the antecedent events for 'falls on the same level'. The worker slipped in 83 (43%) of the incidents, tripped in 35 (18%) of the incidents, lost their balance in 26 (14%), suffered a seizure in 18 (9%), and passed out or fainted in 15 (8%). The remaining 18 cases fell from a chair or stool, fell while snow skiing or ice skating, suffered a fall as a result of a heart attack or as a result of physical combat training. 'Ice, sleet, snow', 'liquids', 'lubricating grease, cutting oils', 'detergent', 'polishes', and 'water' were specifically listed as a secondary source of injury in 14% of all fatal, occupational falls on the same level.

5.2. *US National Health Interview Survey, 1997–1998*
The NHIS asks persons injured at work to attribute the cause of their injury. Table 7 presents weighted and annualised data from 1997 and 1998 for the interviewee-attributed cause of a fall, which resulted in injury at work. The data show that slipping, tripping or stumbling contributed to 64% of all falls, loss of balance to 11%, and jumping or diving to 3%.

5.3. *Swedish Information System Occupational Injuries and Diseases (ISA), 1998*
ISA has a facility to examine events in the causal chain prior to the event to which an injury may be attributed. This has proved to be useful in examining a number of questions using ISA data. Final preceding event data on reported falls in Sweden in 1998 were requested from the Swedish government and are displayed in table 8. The data were obtained from the Swedish Work Environment Authority in Solna, Sweden (Blom 2000). Totals here include 414 cases beyond those published in the official statistics for 1998. These additional cases involved persons in military service, work-related training for no pay, and students.

The 8 224 cases in table 8 do not include cases that involved falling into a tank or pit or jumping to a lower level, which are classified separately from falls to lower levels in ISA. Demonstrating that falls are not the only outcome of concern, there were also 322 cases of overexertion injuries attributed to slipping ($n = 275$) and tripping ($n = 47$), respectively, in 1998.

Table 7. Distribution of causes of all falls at work – NHIS weighted and annualised estimates, 1997/1998.

Causes	Number of events per year	Percentage
Slipping, tripping or stumbling	859 864	64
Loss of balance or dizziness	142 210	11
Jumping or diving	62 044	5
Pushed	35 792	3
Other	235 260	17
Don't know	14 292	1
Total	1 286 180	100

Table 8.   ISA fall related injuries by preceding events for 1998.

Preceding events	Falls on same level		Falls to a lower level	
	Number	Percentage	Number	Percentage
Slipping	2 986	55	636	23
Tripping	1 228	22	145	5
Pushed	123	2	45	2
Fainting, tiredness	77	1	17	0.5
Underlay tipped/rolled/slid	96	2	880	32
Vehicle in motion	25	0.5	23	1
Step into the air	249	5	423	15
Lost grip	59	1	94	3
Other fall	627	11	491	18
Totals	5 470	100	2 754	100

The data from table 8 show a 2:1 ratio of falls on the same level to falls to a lower level. Table 8 also demonstrates that the preceding events for falls on the same level and falls to a lower level were different. While slipping was cited in 44% of the 8 224 falls and tripping in 17% of cases, both made a greater proportionate contribution to the same level falls than to falls to a lower level. Falls to a lower level involved a larger percentage of tipped, rolled or slid underlays than falls on the same level (32% vs. 2%). This category relates to the stability of the interface between standing/walking surfaces and supporting surfaces.

6.   Discussion

6.1. *STF burden overall*

All the data systems permitted an examination of falls on the same level and falls to a lower level. Regarding these two classes of STF-related injury, it may be readily observed that across these data systems (with one exception of the NHIS) non-fatal falls on the same level consistently outnumbered falls to a lower level by a factor of approximately 2 to 1. This is in contrast to fatal falls where falls to a lower level exceed falls on the same level by a factor of 11 to 1. The NHIS presented the curious result of self-reported falling to a lower level exceeding falls on the same level. Contrasted with the other systems, this may suggest a reduced recall for same level falls and may reflect, in part, a reduced likelihood of seeking medical treatment or advice for these sorts of falls compared with falls from elevation.

Non-fatal falls occurred most frequently in the Services sector. However, fall rates for this sector were the lowest of the five sectors considered in table 4 and highest for the mining and construction sectors.

In Sweden, falls to a lower level and on the same level combined accounted for 22% of disabling accidents. In Great Britain, falls accounted for 30% of all injury events reported to the HSE, a sizeable proportion. While exact matches are not available in the USA, the BLS SOII data indicated that the combined STF category (with non-fall slip and trip injuries included) accounted for 21% of all cases involving a day or more away from work. SOII data further show that nearly one-half of disabling sprains and strains and nearly one-half of disabling fractures are attributable to fall-related events.

6.2. *Injuries among older workers*

Increased fall-related mortality and disability with age were noted in figure 1 and table 1, respectively. These findings are consistent with those of Kemmlert and Lundholm (2001), who recently reported that older workers were more likely to report a greater proportion of STF injuries and to experience higher disability durations. The older worker confronts challenges that can impact susceptibility to work-related injury and disability including changes in body dimension, physiological and sensory capacities, and the ability to adaptively respond to physical stressors (Mital 1994, Laflamme and Menckel 1995).

In the general population, it is estimated that one-third of community-dwelling adults aged 65 years or older fall each year (McElbinney *et al.* 1998). Falls are the leading cause of death from injury in those aged 65 years and older and especially those over 75 years old (Sattin 1992). Approximately 9 500 deaths are attributed annually to falls of the elderly in the USA alone. Fall-related mortality in this age group may be related to one of many other contributing factors including co-morbid conditions, *sequelae* of the actual trauma sustained, or nosocomial infections rather than a direct result of the trauma sustained in a fall. As such it is possible that at least some of these injuries are misclassified in existing surveillance systems such as death certifications (Fife 1987).

Kemmlert and Lundholm (2001) reported no differences by age in the exposures contributing to STF injuries and suggested that occupational prevention strategies may not need to differ with age. While it is unclear exactly how the demographic trend of the ageing workforce (Robertson and Tracy 1998, Rantanen 1999) will interact with all these factors, there can be a reasonable expectation that STF-related occupational injuries in older working adults will be one of the expanding challenges in occupational safety in the twenty-first century.

6.3. *Injuries sustained and related disability*

Fatal injuries related to falls often involved head or neck trauma although trauma to internal organs was also a contributing factor. The most disabling non-fatal injuries related to falls were fractures of the pelvis and lower extremities along with some upper extremity fractures and some dislocations. In the USA these most disabling injuries resulted in median absences of as much as 88 days for falls to a lower level and as much as 35 days for falls on the same level (table 5). Courtney and Webster (1999) previously used the data from BLS SOII to examine fractures and reported that falls to a lower level produced the highest median days away from work followed by falls on the same level.

Based on the workers' compensation experience of a large US insurer, the direct insured cost of fall-related occupational morbidity in the USA was estimated at more than US$6 billion annually. Claims for falls to a lower level and falls on the same level combined accounted for 17% of all claims filed and 25% of all claim costs. Leamon and Murphy (1995) studied 278 000 workers' compensation claims from 1989 and 1990 and found that STF represented 16% of all claims and 24% of total claim costs. The present findings suggest that the problem persists.

When costs for the most disabling injuries within a type of fall were considered (table 6), they were found to be as much as 13 times higher than the cost of the average claim for falls to lower levels and as much as 11 times higher for falls on the same level. Higher cost injuries were predominantly fractures and involved the lower extremities, upper extremities and the back.

6.4. *The contribution of slipperiness to STF morbidity and mortality*
Of the data systems consulted only two, the Swedish ISA and the US NHIS, provided any coding to differentiate the contributions of slipping from other types of events in the causal pathway for falls. Descriptive narratives regarding fatal same level falls from the US CFOI were analysed and provided a confirming perspective. Overall in these large data systems there is preliminary evidence that as much as one-half of falls on the same level may be directly attributable to slips. The contribution to falls from height may be more modest based on the ISA data alone at 23% of incidents. The overall ISA finding of slipping contributing to 42% of falling incidents in 1998 is identical to the percentage contribution of slips to falling incidents in Sweden in 1979 as reported by Strandberg (1983).

ISA remains one of the only sources of detailed causal information on large numbers of slipping accidents. However, there are papers in the literature describing studies of accidents that enumerated slips separately from trips and falls. Bentley and Haslam (1998) reported that 42.5% of falls experienced by British mail carriers were the result of slips. Lin and Cohen (1997) reported slips responsible for 13.5% of all injuries. In fast-food outlets Hayes-Lundy *et al.* (1991) reported that 11% of grease burns resulted from slips. Niskanen (1985) reported that slips accounted for 25% of injuries in construction, while McNabb (1994) reported 8% of injuries in petroleum drilling. Shannon and Manning (1980) reported that slipping was the most frequently disabling event resulting in 27% of lost-time injuries in automobile manufacturing.

Ideally, studies and surveillance systems would make use of a detailed method of recording accident data that reflects the sequence of pre-injury or antecedent events. One such system that has been developed and used for this purpose is the Merseyside Accident Information Model (MAIM; Manning 1974). The original concept was described in 1974 and the results of subsequent population-based studies have been published as noted previously (Shannon and Manning 1980, Manning and Ayers 1987, Manning *et al.* 1988, Davies *et al.* 1998, 2001, Manning *et al.* 2000). Currently, MAIM is an intelligent software system designed to collect all available data on injury incidents (Davies and Manning 1994).

The sequence of events, the activities, the human body movements, the environmental objects, the movements and position of the objects and all other contributory factors are addressed as the 'components' in MAIM. However, MAIM's nucleus is the documentation of first unintentional event (first event) in the multi-event incident sequence. For example, a typical event sequence leading to an STF-related injury might include a slip of the foot followed by a loss of balance and then a fall leading to striking the walking surface. Slipping would be recorded as the 'first unintentional event' whereas the fall is recorded as a later event in the timescale of the incident.

6.5. *Limitations of and recommendations for data systems in relation to slipperiness*
Generally, injury surveillance data systems have limitations that should be acknowledged. Owing to resource constraints, many systems have a limited sensitivity for exposure assessment, which proved to be a key factor in the current analysis (Murphy *et al.* 1996, Hagberg *et al.* 1997). The presence of filtering effects at each stage of the reporting process (e.g. a worker's decision to report or an enterprise's decision to report) can also influence what is captured in large-

scale surveillance systems and particularly non-fatal injury systems (Webb *et al.* 1989). Such systems may also be differentially sensitive to various injury and illness types whether planned in design (i.e. hospital trauma surveillance) or unplanned. Finally, most surveillance systems do not collect information on worker experience and socio-economic factors (Cheadle *et al.* 1994, Murphy *et al.* 1996, Hagberg *et al.* 1997, Williams *et al.* 1998, Dassinger *et al.* 1999).

Specific to the issue of STF-related injuries, the majority of data systems consulted could not differentiate slip from non-slip events in fall aetiology. Some like the BLS SOII did document non-fall injuries related to slips but did not isolate the contribution of slipping to fall-related injuries. More often falls were identified according to either the surface the injured fell from or the surface or object onto which they fell. Others like the HSE statistics combined slipping, tripping and falling into one category. Strandberg (1983: 28) commented that 'such single type description models' may lead to 'serious underestimates' of the injury contribution from certain events, agencies and activities. In contrast the Swedish ISA permitted examination of the events preceding the actual fall making possible an estimate of the contribution of slipperiness to fall morbidity. Generally, systems like ISA and MAIM, which incorporate the ability to track multiple events in multi-phase incidents will be more successful in correctly attributing STF its proper burden.

There is also growing interest in the use of free text analysis of narrative descriptions of injury incidents provided within some databases to improve the resolution of injury causation (Sorock *et al.* 1997). Previous barriers of storage for such data and the speed with which they can be analysed are dropping rapidly with the gains in technology in recent years. While the success of such approaches relies on the quality and thoroughness of the narrative data collected, such techniques hold promise not only in better describing exposures but also in understanding the events and event-sequencing that precipitated an injury.

As already recognised, STF-related injuries are the results of fairly complex causal pathways in which ultimate causes are often unknown. Therefore, future surveillance research addressing STF-related occupational injuries should attempt to include to the extent possible (in addition to injury details) a description of:

(1) The environmental conditions in the incident scenario: What was the walking surface? Were there contaminants? What were the visual conditions? Was it outdoors or indoors?
(2) The footwear conditions: What type of footwear and sole material? For how long had the footwear been worn?
(3) The human factors in the incident scenario: Was the victim familiar or unfamiliar with the scene of the incident? What was the purposeful activity the victim was engaged in (e.g. walking, running, carrying, or pulling)? Was there anything unusual about the victim's state of mind or health just preceding the incident (e.g. were they tired, rushed, distracted, medicated)?
(4) The time sequence of the incident scenario.

It is recognised that such details may be difficult to obtain, particularly in cases of fatal injury where there were no witnesses. However, where such improvements can be made, significant gains could be achieved in the understanding of the causes of STF-related injuries.

7. Conclusions

The burden of STF-related occupational injury is substantial comprising between 20 to 40% of disabling occupational injuries in the developed countries studied. This percentage increases if more severely disabling injuries are considered especially for falls to a lower level. These proportions may well reflect an underestimate due to peculiarities of the recording of STF-related injuries discussed previously. The estimated annual US direct cost of fall-related occupational injuries alone was approximately US$6 billion with no evidence of a reduction in losses due to slipping and falling over time. Slipperiness or slips were found to contribute to between 40 and 50% of fall-related injuries. Slipperiness was more often a factor in same level falls (as much as 55% of cases) than in falls to lower levels (as little as 23% of cases).

The evaluation of the burden of slipperiness was hampered by design limitations in many of the data systems utilised. Improvements in the resolution of large-scale injury registries by collecting information on incident sequences should be pursued in order to better define the full scope and contribution of slipperiness as a factor in STF-related injuries. A system like ISA or approach like MAIM could serve as a model for such improvements to other systems. Such improvements would facilitate the allocation of prevention resources towards reduction of first event risk factors such as slipping or tripping that may prove more amenable to prevention than falling, particularly on the same level. There are a number of approaches to prevention including improved housekeeping, improved footwear, better design of walking surfaces, control of contaminants, and better work practices outlined in articles such as Leamon (1992) and more recently Bentley and Haslam (2001a, b).

Examining the STF burden almost a decade earlier, Leamon and Murphy (1995) concluded that 'based on the frequency and costs to industry and workers, prevention of falls should be given a high priority'. Despite the evidence presented here, elsewhere in the literature, and the existence of sound literature on prevention, the problem appears to be as pervasive as ever.

Acknowledgements

The authors would like to thank the following individuals who assisted in the acquisition of data from various surveillance systems for the study: Kjell Blom, John Cotnam, Raoul Grönqvist, Pamela Hixon, Lawrence Jackson, Simon Matz, Margaret Warner and Helen Wellman. The authors would also like to thank our reviewers for their constructive criticisms of earlier drafts of the manuscript: Wen Chang, Raoul Gronqvist, Lynn Jenkins, Tom Leamon, Mary Lesch, Suzanne Marsh, Timothy Pizatella, Mark Redfern, William Shaw, Gordon Smith, Lennart Strandberg, and all of our Measurement of Slipperiness Symposium colleagues.

References

ANDERSSON, R. and LAGERLOF, E. 1983, Accident data in the new Swedish information system on occupational injuries, *Ergonomics,* **26,** 33–42.

BENTLEY, T. A. and HASLAM, R. A. 1998, Slip, trip and fall accidents occurring during the delivery of mail, *Ergonomics,* **41,** 1859–1872.

BENTLEY, T. A. and HASLAM, R. A. 2001a, A comparison of safety practices used by managers of high and low accident rate postal delivery offices, *Safety Science,* **37**(1), 19–37.

BENTLEY, T. A. and HASLAM, R. A. 2001b, Identification of risk factors and countermeasures for slip, trip and fall accidents during the delivery of mail, *Applied Ergonomics,* **32,** 127–134.

BLOM, K. 2000, Personal communication.

CHEADLE, A., FRANKLIN, G., WOLFHAGEN, C., SAVARINO, J., LIU, P. Y., SALLEY, C. and WEAVER, M. 1994, Factors influencing the duration of work-related disability: a population-based study of Washington State workers' compensation, *American Journal of Public Health*, **84**, 190–196.

COURTNEY, T. K. and WEBSTER, B. S. 1999, Disabling occupational morbidity in the United States: an alternative way of seeing the Bureau of Labour Statistics' data, *Journal of Occupational and Environmental Medicine*, **41**, 60–69.

COURTNEY, T. K. and WEBSTER, B. S. 2001, Antecedent factors and disabling occupational morbidity—insights from the new BLS data, *American Industrial Hygiene Association Journal*, **62**, 622–632.

DASSINGER, L. K., KRAUSE, N., DEEGAN, L. J., BRAND, R. J. and RUDOLPH, L. 1999, Duration of work disability after low back injury: a comparison of administrative and self-reported outcomes, *American Journal of Industrial Medicine*, **35**, 619–631.

DAVIES, J. C. and MANNING, D. P. 1994, MAIM: the concept and construction of intelligent software, *Safety Science*, **17**, 207–218.

DAVIES, J. C., STEVENS, G. and MANNING, D. P. 1998, Understanding accident mechanisms: an analysis of the components of 2516 accidents collected in a MAIM database, *Safety Science*, **29**, 25–58.

DAVIES, J. C., STEVENS, G. and MANNING, D. P. 2001, An investigation of underfoot accidents in a MAIM database, *Applied Ergonomics*, **32**, 141–147.

DEMPSEY, P. G. and HASHEMI, L. 1999, Analysis of workers' compensation claims associated with manual materials handling, *Ergonomics*, **42**, 183–195.

FIFE, D. 1987, Injuries and deaths among elderly persons, *American Journal of Epidemiology*, **126**, 936–941.

FINGERHUT, L. A., COX, C. S. and WARNER, M. 1998, *International Comparative Analysis of Injury Mortality: Findings from the ICE on Injury Statistics*, Advance data from vital and health statistics, no. 303 (Hyattsville, MD: National Center for Health Statistics).

GRÖNQVIST, R. and ROINE, J. 1993, Serious occupational accidents caused by slipping, in R. Nielsen and R. Jorgensen (eds), *Advances in Industrial Ergonomics and Safety V* (London: Taylor & Francis), 515–519.

HAGBERG, M., CHRISTIANI, D., COURTNEY, T. K., HALPERIN, W., LEAMON, T. B. and SMITH, T. J. 1997, Conceptual and definitional issues in occupational injury epidemiology, *American Journal of Industrial Medicine*, **32**, 106–115.

HAYES-LUNDY, C., WARD, R. S., SAFFLE, J. R., REDDY, R., WARDEN, G. D. and SCHNEBLY, W. A. 1991, Grease burns at fast-food restaurants—adolescents at risk, *Journal of Burn Care and Rehabilitation*, **12**, 203–208.

HEALTH and SAFETY EXECUTIVE (HSE) 1999, *Health and Safety Statistics 1998/9* (Sudbury, UK: HSE Books).

JACKSON, L. and HIXON, P. 2000, Personal communication.

KEMMLERT, K. and LUNDHOLM, L. 1998, Slips, trips and falls in different work groups with reference to age, *Safety Science*, **28**, 59–75.

KEMMLERT, K. and LUNDHOLM, L. 2001, Slips, trips and falls in different work groups with reference to age and from a preventive perspective, *Safety Science*, **32**, 149–153.

LAFLAMME, L. and MENCKEL, E. 1995, Aging and occupational accidents: a review of the literature of the last three decades, *Safety Science*, **21**, 145–161.

LEAMON, T. B. 1992, The reduction of slip and fall injuries: Part 1—Guidelines for the practitioner, *International Journal of Industrial Ergonomics*, **10**, 23–27.

LEAMON, T. B. and MURPHY, P. L. 1995, Occupational slips and falls: more than a trivial problem, *Ergonomics*, **38**, 487–498.

LIN, L. J. and COHEN, H. H. 1997, Accidents in the trucking industry, *International Journal of Industrial Ergonomics*, **20**, 287–300.

MANNING, D. P. 1974, An accident model, *Occupational Safety and Health*, January, 14–16.

MANNING, D. P. 1983, Deaths and injuries caused by slipping, tripping and falling, *Ergonomics*, **26**, 3–9.

MANNING, D. P. and AYERS, I. M. 1987, Disability resulting from underfoot first events, *Journal of the Society of Occupational Medicine*, **37**, 39–41

Manning, D. P., Davies, J. C., Kemp, G. J. and Frostick, S. P. 2000, The Merseyside Accident Information Model (MAIM) can reveal components of accidents that lead to attendance at fracture clinics and cause disability: a new approach to accident prevention, *Safety Science*, **36**, 151–161.

Manning, D. P., Ayers, I., Jones, C., Bruce, M. and Cohen, K. 1988, The incidence of underfoot accidents during 1985 in a working population of 10,000 Merseyside people, *Journal of Occupational Accidents*, **10**, 121–130.

McElbinney, J., Koval, K. J. and Zucherman, J. D. 1998, Falls in the elderly, *Archives of the American Academy of Orthopaedic Surgeons*, **2**(1), 60–65.

McNabb, S. J., Ratard, R. C., Horan, J. M. and Farley, T. A. 1994, Injuries to international petroleum drilling workers, *Journal of Occupational Medicine*, **36**, 627–630.

Mital, A. 1994, Issues and concerns in accommodating the elderly in the workplace, *Journal of Occupational Rehabilitation*, **4**, 253–267.

Murphy, P. L. and Courtney, T. K. 2000, Low back pain disability: relative costs by antecedent and industry group, *American Journal of Industrial Medicine*, **37**, 558–571.

Murphy, P. L., Sorock, G. S., Courtney, T. K., Webster, B. S. and Leamon, T. B. 1996, Injury and illness in the American workplace: A comparison of data sources, *American Journal of Industrial Medicine*, **30**, 130–141.

National Safety Council (NSC) 1995, *International Accident Facts* (Itasca, IL: NSC).

National Safety Council (NSC) 1998, *Accident Facts* (Itasca, IL: NSC).

Niskanen, T. 1985, Accidents and minor accidents of the musculoskeletal system in heavy (Concrete Reinforcement Work) and light (painting) construction work, *Journal of Occupational Accidents*, **7**(1), 17–32.

Rantanen, J. 1999, Research challenges arising from changes in worklife, *Scandinavian Journal of Work, Environment & Health*, **25**(6, special issue), 473–483.

Robertson, A. and Tracy, C. S. 1998, Health and productivity of older workers, *Scandinavian Journal of Work, Environment & Health*, **24**(2), 85–97.

Sattin, R. W. 1992, Falls among older persons: a public health perspective, *Annual Reviews of Public Health*, **13**, 489–508.

Shannon, H. S. and Manning, D. P. 1980, Differences between lost-time and non-lost-time industrial accidents, *Journal of Occupational Accidents*, **2**, 265–272.

Sorock, G. S., Smith, G., Reeve, G., Dement, J., Stout, N., Layne, L. and Pastula, S. 1997, Three perspectives on work-related injury surveillance systems, *American Journal of Industrial Medicine*, **32**, 116–128.

Strandberg, L. 1983, On accident analysis and slip-resistance measurement, *Ergonomics*, **26**, 11–32.

Swedish Work Environment Authority and Statistics Sweden 2000, *Occupational Diseases and Occupational Accidents 1998* (Stockholm: SWEA).

US Centres For Disease Control And Prevention 1998, Surveillance for non-fatal occupational injuries treated in hospital emergency departments, United States 1996, *Morbidity Mortality Weekly Report*, **47**, 302–306.

US Department of Labour, Bureau of Labour Statistics (USDOL-BLS) 1992, *Occupational Injury and Illness Classification Manual* (Washington, DC: US Government Printing Office).

US Department of Labour, Bureau of Labour Statistics (USDOL-BLS) 1997, BLS *Handbook of Methods*, Bulletin 2490: 70–88 (Washington, DC: US Government Printing Office).

US Public Health Service and Health Care Financing Administration 1991, *International Classification of Diseases, 9th Revision, Clinical Modification* (Washington, DC: US Public Health Service).

Warner, M., Barnes, P. M. and Fingerhut, L. A. 2000, *Injury and Poisoning Episodes and Conditions: National Health Interview Survey, 1997*. Vital and Health Statistics, Series 10, no. 303 (Hyattsville, MD: National Centre for Health Statistics).

Webb, G. R, Redman, S., Wilkinson, C. and Sanson-Fisher, R. W. 1989, Filtering effects in reporting work injuries, *Accident Analysis & Prevention*, **21**, 115–123.

Webster, B. S. and Snook, S. H. 1990, The cost of compensable low back pain, *Journal of Occupational Medicine*, **32**, 13–15.

WEBSTER, B. S. and SNOOK, S. H. 1994, The cost of 1989 workers' compensation low back pain claims, *Spine,* **19,** 1111–1116.

WILLIAMS, D. A., FEUERSTEIN, M., DURBIN, D. and PEZZULLO, J. 1998, Health care and indemnity costs across the natural history of disability in occupational low back pain, *Spine,* **23,** 2329–2336.

CHAPTER 3

Biomechanics of slips

Mark S. Redfern†*, Rakié Cham†, Krystyna Gielo-Perczak‡, Raoul Grönqvist§,
Mikko Hirvonen§, Håkan Lanshammar¶, Mark Marpet††, Clive Yi-Chung Pai‡‡
and Christopher Powers§§

†Department of Bioengineering, University of Pittsburgh, 200 Lothrop Street,
Pittsburgh, PA 15213, USA

‡Liberty Mutual Research Center for Safety and Health, Hopkinton, MA, USA

§Finnish Institute of Occupational Health, Finnish Institute of Occupational
Health, Department of Physics, FIN-00250 Helsinki, Finland

¶Information Technology Department of Systems & Control, Uppsala University, Uppsala, Sweden

††St. John's University, New York, NY, USA

‡‡Department of Physical Therapy, University of Illinois at Chicago, Chicago,
IL, USA

§§Department of Biokinesiology, University of Southern California, Los Angeles,
CA

Keywords: Biomechanics; Slipping; Gait; Postural control; Ground reaction
forces.

The biomechanics of slips are an important component in the prevention of fall-
related injuries. The purpose of this paper is to review the available literature on
the biomechanics of gait relevant to slips. This knowledge can be used to develop
slip resistance testing methodologies and to determine critical differences in
human behaviour between slips leading to recovery and those resulting in falls.
Ground reaction forces at the shoe-floor interface have been extensively studied
and are probably the most critical biomechanical factor in slips. The ratio of the
shear to normal foot forces generated during gait, known as the required
coefficient of friction (RCOF) during normal locomotion on dry surfaces or
'friction used/achievable' during slips, has been one biomechanical variable most
closely associated with the measured frictional properties of the shoe/floor
interface (usually the coefficient of friction or COF). Other biomechanical factors
that also play an important role are the kinematics of the foot at heel contact and
human responses to slipping perturbations, often evident in the moments
generated at the lower extremity joints and postural adaptations. In addition, it
must be realised that the biomechanics are dependent upon the capabilities of the
postural control system, the mental set of the individual, and the perception of the
environment, particularly, the danger of slipping. The focus of this paper is to
review what is known regarding the kinematics and kinetics of walking on
surfaces under a variety of environmental conditions. Finally, we discuss future
biomechanical research needs to help to improve walkway-friction measurements
and safety.

---

*Author for correspondence. e-mail: mredfern@pitt.edu

1. Introduction

Pedestrian accidents on walkways continue to be a very serious problem. An analysis of data in 1986 found that the costs of pedestrian accidents was second in magnitude only to automobile accidents and that falls were the leading cause of accidental death in senior citizens (Rice *et al.* 1989). Pedestrian-fall accidents have been the second largest generator of unintentional workplace fatalities (Leamon and Murphy 1995), and accounted for nearly 11% and 20%, respectively, of all fatal and non-fatal (involving lost work days) occupational injuries in the USA in 1996 (US Department of Labor, Bureau of Labor Statistics 1997, 1998).

Falls precipitated by slipping are of major concern (Courtney *et al.* 2001). Lloyd and Stevenson (1992) reported that slips and trips cause 67% and 32% of falls sustained by the elderly and young, respectively. The magnitude of the problem is probably greater than suggested by the statistics above, as muscular strain or back pain illnesses resulting from slip and recovery incidents are usually not reported as related to slips (Manning and Shannon 1981, Troup *et al.* 1981, Anderson and Lagerlof 1983).

Understanding what causes slip-precipitated pedestrian accidents is challenging because of the multiple, interacting environmental and human factors involved. Among the environmental factors are properties of the walking-surface (such as surface roughness, compliance, topography, as well as the properties of adjacent areas and contaminants) and/or shoe or foot (e.g. material properties, tread and wear). Other environmental factors include lighting and contrast levels, and climatic factors such as ice and snow. Human factors include gait, expectation, the health of the sensory systems (i.e. vision, proprioception, somatosensation, and vestibular) and the health of the neuromuscular system.

One fundamental principal in determining the slip propensity of a given situation is the relationship between the friction required by the pedestrian for the manoeuvre being conducted (*Required friction*) compared with the friction available at the walkway/shoe interface (*Available friction*). Theoretically, as long as the available friction exceeds the required friction, the pedestrian will not slip. There are a variety of pedestrian gaits, e.g. level walking, load carrying, walking up ramps, that have different levels of required friction to prevent slip. Thus, biomechanical analysis of gait is potentially a valuable tool in the reduction of slip-induced fall accidents because it can illuminate the conditions that may be hazardous to pedestrians. Furthermore, biomechanical analysis of gait can be an important input into the setting of available friction thresholds to determine whether or not a shoe, walkway surface, or combination of the two will—or will not—be slip resistant (Marpet 1996).

The utility of biomechanics in the measurement of slipperiness goes beyond the matter of determining required friction. Biomechanics can be employed to 'tune' tribometric instruments (here, walkway-surface friction testing instruments) to reflect the friction situation facing pedestrians. The friction model taught in high-school and college physics classes (the Amontons-Coulomb model) assumes that friction is solely a material property of the interface materials, independent of contact area, pressure, temperature, velocity, etc. (Amontons 1699, Coulomb 1781). This model, true for friction between two rigid bodies, is an oversimplification with respect to pedestrian friction. James (1980, 1983) showed that the materials commonly used in shoe-bottom construction do not follow the assumptions of Amontons-Coulomb. Given that, devices that measure the frictional properties of the shoe-floor interface will give more meaningful results if testing conditions, such as pressure, velocity, contact time and so forth, mimic those of pedestrian gait. The need for such *biofidelity* in friction testing has been recognised for some time (Proctor and Coleman 1988). At present, no walkway tribometer has operational characteristics mimicking human gait. New methods do attempt to obtain biomechanically relevant measures of slip resistance (Grönqvist *et al.* 1989, Wilson 1990, Redfern and Bidanda 1994). Gait parameters describing foot dynamics during actual slip events under varying environmental conditions (floor and contaminant) are important in the further development of these devices.

Another aspect of slip-precipitated falls that incorporates biomechanics is the capability of the human postural control system that is used to maintain balance and recover from perturbations. Balance recovery during slips involves neuromuscular control, biomechanics, and their interaction with the environment. Therefore, to understand the impact of the environment on the potential of slips and falls, some knowledge of the human reactions to slip and the capability to recover from slip perturbations must be taken into consideration. Factors to consider in balance recovery and control are anatomical (e.g. foot geometry, body mass and its distribution, or segment length and height), physiological (e.g. strength, rate of muscle force rise, or gains and delays of feedback control), or perhaps cognitive and behavioural constraints (e.g. reaction time, attention, or fear of falling). Each of these constraints has a different impact on the ability to recover balance, which can be assessed in terms of kinematics and kinetics of the performance.

Thus, the main goal of this paper is to review the relevant literature investigating gait biomechanics and postural control while walking on surfaces of varying slipperiness. The focus of this literature review will be on the kinematics and kinetics of walking on surfaces of different inclination under a variety of environmental conditions. A second goal is to define the specific research directions and goals needed to determine which gait parameters are important in supporting improved walkway-friction measurements and safety.

2. Biomechanics of locomotion without slipping

This section describes the biomechanics of common locomotor activities without slipping. In all cases, the biomechanical descriptions include ground reaction forces, kinematics and moments generated at the lower extremity joints. Normal gait is first described for walking on a level surface, and then on inclined surfaces. The figures for level walking and inclined surface walking (when available) have been combined for ease of comparison (figures 1, 3–6). The next section describes the biomechanics of another locomotor activity known to be associated with slips and falls, namely, ascending and descending stairs. Finally, the influence of load carrying during walking on the biomechanics of gait is considered.

2.1. *Walking on level surfaces*
2.1.1. *Ground reaction forces:* The force interactions between the shoe and floor are probably the most critical biomechanical parameters in slips and falls. If the shear forces generated during a particular step exceed the frictional capabilities of the shoe/floor interface, then a slip is inevitable. Thus, an understanding of the forces at the shoe/floor interface is important. A number of researchers have examined foot forces, often termed ground reaction forces (GRF), during normal gait on a level surface (Strandberg and Lanshammar 1981, Perkins and Wilson 1983, Strandberg 1983, Winter 1991, Redfern and Dipasquale 1997, Cham and Redfern 2002a). (See

Table 1. Foot force parameters for normal walking on level surfaces (0°) and inclined surfaces (5° and 10°).

Variable: Mean (SD)	0°	5°	10°
First peak of normal forces (body weight N kg^{-1})	10.92 (1.42)[†]	12.15 (1.41)[†]	13.33 (1.52)[†]
Peak shear forces (body weight N kg^{-1})	1.77 (0.61)[†]	2.94 (0.56)[†]	4.06 (0.81)[†]
Timing of peak RCOF (% stance or ms)	16.5 (2.4)[†]% 91 (25)[††]ms	18.1 (3.6)[†]	19.2 (4.6)[†]
Timing of first peak of normal forces (% stance)	24.5 (5.2)[†]	21.4 (4.3)[†]	18.6 (5.4)[†]
Timing of peak shear forces (% stance)	19.0 (3.1)[†]	19.5 (2.6)[†]	19.0 (4.6)[†]
Peak RCOF*	0.17 (0.04)[††] 0.18 (0.05)[†] 0.18 (0.06)[‡] 0.20[¶] 0.22[§]	0.26 (0.03)[†]	0.32 (0.05)[†] 0.33 (0.04)[‡]

*Required Coefficient of Friction, defined in section 2.2.1.

From [†]Cham and Redfern [vinyl floors] (2002a), [‡]Hanson *et al.* (1999), [§]Perkins (1978), [¶]Redfern and DiPasquale (1977), [††]Strandberg (1983) ['grip' trials].

table 1 for a list of critical GRF parameters.) The normal forces (perpendicular to the walking surface) are typically characterised by two peaks (solid line in figure 1). The first peak occurs at the end of the loading phase (about 25% into stance) as full body weight is transferred to the supporting foot, while the second peak occurs later in stance just prior to the beginning of the toe-off phase. The anterior-posterior shear forces exhibit a biphasic, symmetrical shape with the first major peak in the forward direction attributed to the loading dynamics, while the second maximum in the rearward direction happens as the heel rotates off the floor pushing back the toes to start the toe-off phase. The first peak in shear force is considered to be the critical one with respect to slips resulting in falls. It occurs at about 19% into the stance phase, which is 90 to 150 ms after heel contact depending on stance duration.

The forces occurring shortly after heel contact have been thought to play an important role in slips and falls. Small spikes in the GRFs have been recorded (Perkins 1978, Lanshammar and Strandberg 1981, Whittle 1999) (figure 2). The first spike immediately after heel contact tends to be in the anterior or forward direction. This spike is not always evident, perhaps because of measurement error due to very low force levels or to low sampling rates. However, those who have recorded these forces have attributed them to the movement of the heel as it impacts the floor and transfers momentum to the ground (Perkins 1978, Whittle 1999). Another peak in the shear GRF that may be more important to slips and falls occurs a little later and is in the rearward direction (Perkins 1978, Lanshammar and Strandberg 1981, Strandberg 1983). A more careful examination of the heel kinematics by

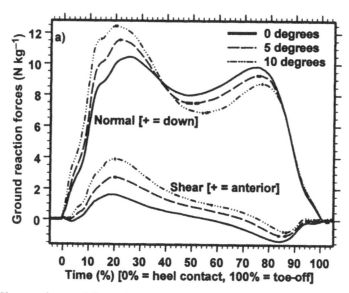

Figure 1.   Shear and normal forces for walking along horizontal surfaces and down inclined surfaces (from Cham and Redfern 2002a).

Lanshammar and Strandberg suggested that this spike is due to rearward movement of the heel during the early loading phase. As the foot rotates down on the floor and reaches foot-flat position (at about 15% into stance), it creates another broader spike especially evident in the normal GRF (Whittle 1999).

Since the shear forces are highest near the heel contact and push-off phase (Redfern and DiPasquale 1997), these are the points where slips most often occur. Heel contact is the critical phase where slips can result in falls (Strandberg 1983, Rhoades and Miller 1988, Lloyd and Stevenson 1992, Redfern and Bloswick 1995, Hanson *et al.* 1999). Thus, the forces occurring at heel contact are of critical importance in determining if the frictional capabilities of the shoe/floor interface will be sufficient to prevent slips.

One GRF measure that has been used to quantify and understand the biomechanics of slips has been the ratio of shear to normal GRF components. During normal locomotion on dry surfaces, i.e. no-slip conditions, this ratio has been described as the 'required coefficient of friction' (RCOF) (Redfern and Andres 1984, Rhoades and Miller 1988, Grönqvist *et al.* 2001a). As a result of the normal and shear force profiles described earlier, the RCOF has a peak value occurring at about the same time as the peak shear force. This peak value is about 0.20 (McVay and Redfern 1994). The peak RCOF has been suggested to predict slip potentials for various gait activities (Redfern and Andres 1984, Love and Bloswick 1988, Buczek *et al.* 1990, McVay and Redfern 1994, Buczek and Banks 1996).

2.1.2. *Kinematics of walking*:   Walking speed is an obvious characteristic that will impact slip potential. Laboratory measurements of self-chosen gait speeds have ranged from 0.97 m s^{-1} (Redfern and DiPasquale 1997) to 1.51 m s^{-1} (Murray *et al.* 1967) for level surfaces. Sun *et al.* (1996) reported walking speeds of 1.1 – 1.2 m s^{-1} on a considerably large number of subjects in a natural urban setting. Step length

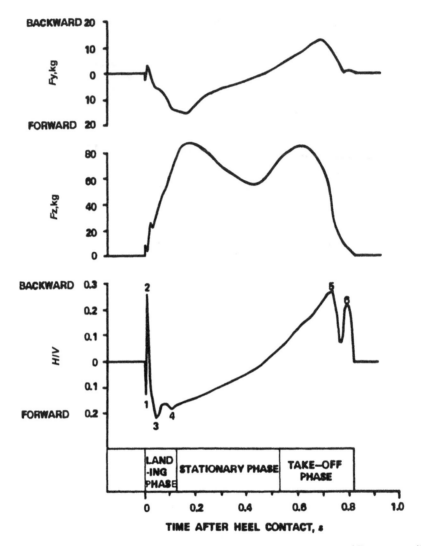

Figure 2.   Typical ground reaction forces (F_y = anterior-posterior shear and F_z = normal) and
required coefficient of friction (H/V) during stance. Note that peak 1 is caused by the
forward force of impact of the heel onto the force plate. Peak 2 is a result of a backward force
exerted on the heel after contact during the early landing phase. Peaks 3 and 4, often
recorded as one broad spike, are caused by the main forward force, which retards the motion
of the foot. Finally, peaks 5 and 6 are recorded during the push-off phase, with the toes in
contact with the force plate, pushing in the backward direction (from Perkins 1978).

theoretically has an important effect on slip potential. The influence of step length on
slip potential was explored using a static model by Grieve (1983). As the step length
is increased, the ratio of shear to normal forces at heel contact would change,
resulting in a greater shear force during the initial portion of the step. Thus, reducing
step length is one method that can reduce the slip potential when walking.

The kinematics of the heel as it comes in contact with the floor is believed to have a
role in the potential for slips and falls (Redfern and Bidanda 1994). Recordings of heel

movements have shown that the heel rapidly decelerates just prior to heel contact, then there is a slight sliding motion along the surface at impact (Strandberg and Lanshammar 1981, Perkins and Wilson 1983, Cham and Redfern 2002a). (See solid line in figure 3a for level walking.) The patterns of sliding during this time can be variable. In general, studies have shown that the heel velocity is forward immediately upon impact, then either coming to a stop or reversing sliding direction before coming

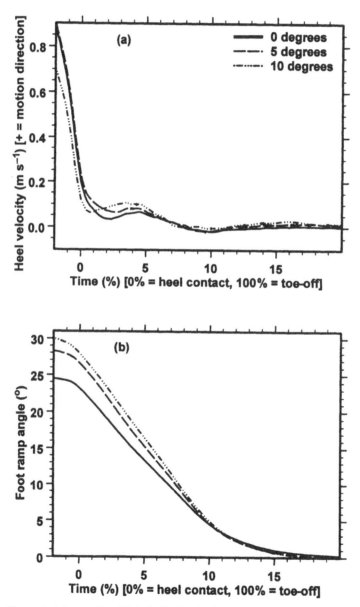

Figure 3.   Characteristic profile of (a) the heel velocity, and (b) the foot-ramp angle (0°, 5°, 10°), averaged across trials conducted on dry vinyl tile flooring. (Time is truncated at 20% of the stance for a more detailed view of heel contact dynamics.) Adapted from Cham and Redfern 2002a.

to a stop. However, Cham and Redfern (2002a) reported a significant number of walking trials where the heel's impact velocity in the antero-posterior direction was negative (evident in the standard deviation associated with the heel velocity in table 2), i.e. the heel was moving in the rearward direction at the instant of contact. In all reported cases, this rapid heel motion ended shortly after heel contact and the heel came to a complete stop, while the foot continued to rotate down on the floor from about 23° for level walking, reaching a foot flat position about 15% into stance.

Joint angles were investigated by Murray *et al.* (1967), Winter (1991) (level walking), Redfern and DiPasquale (1997), and Cham and Redfern (2002a). In general, the overall profiles of joint angles were in agreement across studies. (See solid line in figure 4 for level walking angle profiles.) At heel contact, the ankle is in slight dorsiflexion but rapidly reaches its peak plantar flexion angle (around 10% of stance phase) as the foot rotates down onto the floor. Toe-off phase begins about 80% into stance when the heel comes off the floor and the ankle again goes into plantar flexion. During the first 30% of the stance, there is an increased flexion of the knee, caused mainly by the forward rotation of the shank. During the last phase of the stance (>60–80%), characterised by the movement of the body's centre of gravity past the single leg base of support, the knee flexes again as the subject prepares for the heel contact of the other foot (second half of the non-supporting leg's swing phase) and toe-off of the supporting foot. The hip angle profile reflects the changes in the upper leg orientation, i.e. only small variations in torso orientation (a few degrees) have been recorded. For most of the stance duration the hip is in extension due to the continuous forward rotation of the upper leg. However, at the end of the stance phase, the subject prepares for the swing phase by rotating the foot off the floor, flexing the knee and the hip (via rearward rotation of the upper leg).

2.1.3. *Joint moments*: One biomechanical measure that sometimes has been overlooked with regards to slips and falls is joint moments. Moments at the ankle,

Table 2. Foot and heel kinematic parameters for normal walking on level dry surfaces (0°) and inclinded surfaces (5° and 10°).

Variable: Mean (SD)	0°	5°	10°
Heel velocity in the direction of motion at HC (m s^{-1})	0.19 (0.39)[†] 0.14 (0.27)–0.68 (0.52)[¶] 1.03 (0.16)[††] during 60 ms prior to heel contact 0.3–2.75[‡‡]	0.25 (0.42)[†]	0.13 (0.32)[†]
Foot angular velocity at HC (° s^{-1})	223.8 (98.4)[†]	251.7 (111.9)[†]	292.9(86.9)[†]
Heel contact angle (°)	23.5 (3.7)[†] ≈30[‡] 10–30[§] 22 (5.3)[¶] 32 (4)[††] during 60 ms prior to heel contact	26.4 (3.5)[†]	26.9 (4.9)[†]

From [†]Cham and Redfern (2002a) [vinyl floors], [‡]Leamon and Son (1989), [§]Perkins (1978), [¶]Strandberg (1983) ['grip' trials and heel velocity averaged within subjects], [††]Grönqvist (1999), [‡‡]Morach (1993).

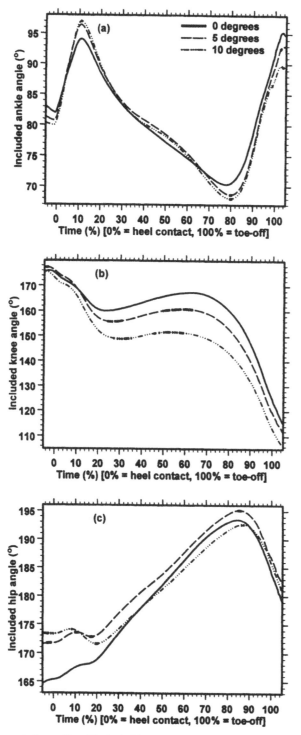

Figure 4.   Characteristic profile of included joint angles during gait on the vinyl tile floor (0°, 5°, 10°), averaged across all trials: (a) ankle, (b) knee, and (c) hip (adapted from Cham and Redfern 2002a).

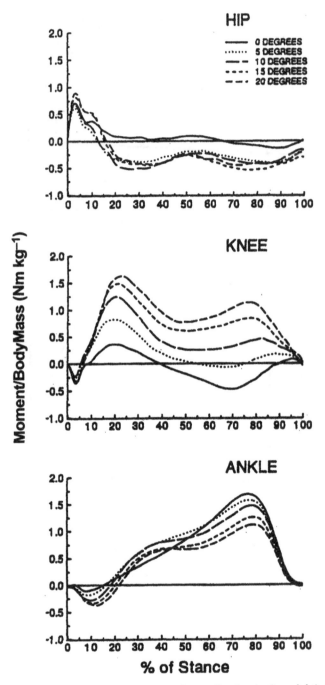

Figure 5.   Characteristic profile of joint moments (normalised to body weight) during gait on level and inclined surfaces (from Redfern and DiPasquale 1997).

knee and hip are required to maintain upright walking. These moments are related to the strength level required to walk, and actions to recover from a slip if necessary.

The moments at the ankle, knee and hip joints have been calculated for level walking by a number of researchers for a variety of subject populations (Winter 1991). For a healthy population, the moments for level walking are shown in the solid line of figure 5. Note the biphasic (extension-flexion) nature of the knee moment and increasing plantar flexion moment at the ankle.

2.2. *Walking on an inclined surface*

Walking on an inclined surface, i.e. ramp, changes the characteristics of gait, and therefore the potential for slipping. Walking down a ramp increases the risk of slips and falls much more than walking up because of the increased shear forces generated and the reduced ability to recover should a slip occur. For example, Haslam and Bentley (1999) reported that falls in postal workers occurred 30% of the time walking down a sloped drive compared to 2% walking up. This section presents the changes that occur in the biomechanics of gait when walking down a sloped surface, which increases the risk of slips and falls.

2.2.1. *Ground reaction forces on inclined surfaces*:

Changes in the inclination of the floor, i.e. increasing ramp angle, are associated with changes in the ground reaction forces (Redfern and DiPasquale 1997, Cham and Redfern 2002a) (figure 1). For example, shear forces for level walking reach a maximum of about 1.5 to 1.8 N kg^{-1} (normalised to body weight). However, walking down a ramp increases this peak shear by about 61% for a 5° ramp angle and 128% for a 10° ramp angle. The timing of these peak shear forces on ramps appears to be the same as walking on level surfaces. The normal forces also are affected by inclination angle, with an increase in the peak force of about 1 N kg^{-1} for a 5° increase (table 1). The peak normal force occurs earlier on inclined surfaces, leading to a time difference of about 5% between the peak of shear and normal foot forces on level surfaces, and an almost in-phase foot force when the ramp angle was increased to 10°. All these changes affect the RCOF, with the peak RCOF increasing with inclination of the surface. For example, walking down a 20° ramp creates an increase in the peak RCOF from the level value of 0.18 to 0.45 (figure 6). Table 1 shows the increase in the GRFs and peak RCOFs as ramp angle is changed. Ramp angle also has an effect on RCOF for walking up (McVay and Redfern 1994); however, the peak RCOF occurs towards the end of the push-off phase when slips do not usually result in falls.

2.2.2. *Kinematics on inclined surfaces*:

Walking down an inclined surface has been found to have an effect on some kinematics variables, but not on others. For example, natural gait velocity was not found to be significantly different when walking down a ramp compared to walking on a level surface (Sun *et al.* 1996, Redfern and DiPasquale 1997). However, step length was reduced as ramp angle was increased. More specifically, Redfern and DiPasquale (1997) reported a step length of 0.54 m and 0.48 m for level walking and during descent of a 20° ramp, respectively, while Sun *et al.* (1996) found these values equal to about 0.62 and 0.57 m for nearly-horizontal (2°) surfaces and 9° ramps, respectively.

The kinematics of the foot upon heel impact during the descent of inclined surfaces are similar to those for level walking, especially the profile of the heel's linear velocity along the floor surface and the sliding patterns of the heel along with the slip-distance from heel contact. Other variables were slightly more affected by the walkway inclination including foot-floor angle and foot angular velocities recorded

M. S. Redfern et al.

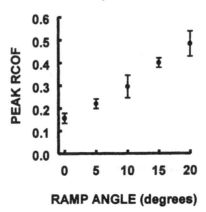

Figure 6.   Peak required coefficient of friction (RCOF) as a function of ramp angle (adapted from Redfern and DiPasquale 1997).

at heel contact (table 2). The foot reaches foot-flat position at about the same time in the gait cycle (15% of stance) as when walking on a level surface. Surface inclination angle has an effect on joint angles during gait (figure 4). While there are minor changes at the hip and ankle, the included knee angle is most affected (Redfern and DiPasquale 1997, Cham and Redfern 2002a).

2.2.3. *Joint moments on inclined surfaces*:   The inclination of a surface can have a significant effect on the moments at the lower extremity joints. Figure 5 shows typical moments during gait for walking on a level surface and the impact of ramp angle (Redfern and DiPasquale 1997). As depicted in figure 5, the resulting moments at ankle, knee and hip were found to change as a function of ramp angle, with the knee moment being the most affected by ramp angle. Redfern and DiPasquale (1997) reported an increase from a mean of 0.4 Nm kg^{-1} when walking on a horizontal surface to 1.7 Nm kg^{-1} when descending a 20° ramp.

2.3. *Walking on stairs*
From a functional standpoint, stair ambulation is a much more challenging task when compared to level gait or walking on ramps. While negotiating stairs, the body is carried in both a vertical and forward direction, which results in joint motion and muscular demands that differ significantly from walking. Walking on stairs not only challenges the strength and range of motion limits of the lower extremity, but also requires substantial balance and muscle co-ordination as well. Owing to the vertical nature of stair ambulation, a slip on stairs can result in a catastrophic event resulting in serious injury. A thorough understanding of the biomechanics during stair negotiation is important for understanding how slips and falls can be prevented during this high demand task.

2.3.1. *Ground reaction forces during stair ambulation*
   2.3.1.1. *Ascending stairs*:   As with level walking, the vertical GRF during stair ascent demonstrates two distinctive peaks, one during weight acceptance (i.e. 30% of stance) and the other during late stance (McFayden and Winter 1988). In contrast to level walking, however, the second peak tends to be slightly greater than the first (figure 7(a)). The higher second peak illustrates the increased force applied to the

floor through strong contraction of the plantarflexors as the body is being elevated to the next step.

Ground reaction force patterns in the anterior-posterior and medial-lateral directions are similar to those during level walking. At weight acceptance, there is an anterior shear force acting on the floor while there is a posterior shear force acting on the floor at toe-off (McFayden and Winter 1988). As shown in figure 7(b), the anterior shear force tends to be somewhat greater in magnitude than the posterior shear force (10% vs 5% body-weight). There also is a lateral shear force acting on the floor that is fairly consistent throughout stance reaching a maximum value of approximately 5% body-weight (figure 7(c)).

As described earlier, the ratio of the resultant shear forces to the vertical (or normal) force has been described as the required coefficient of friction or RCOF. During stair ascent, the RCOF during weight acceptance has been observed to be consistent with values reported for level walking (0.21); however, the RCOF during toe-off appears to be somewhat higher (0.39) (figure 7(d)). Such data suggests that a slip would be more likely to occur in the posterior direction during late stance as the body is being elevated.

2.3.1.2. Descending stairs: The vertical ground reaction force pattern during stair descent varies significantly compared to that of stair ascent (McFayden and Winter 1988). The peak vertical ground reaction force (which also occurs during

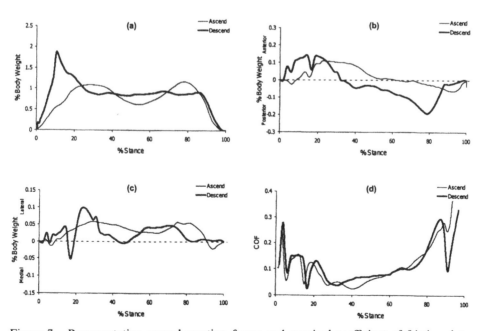

Figure 7. Representative ground reaction forces and required coefficient of friction data obtained during stair ascent and descent in a healthy adult subject (AMTI force plate, 2400 Hz): (a) vertical ground reaction force, (b) anterior-posterior ground reaction force, (c) medial-lateral ground reaction force, and (d) required coefficient of friction (resultant shear force/normal force). Coefficient of friction data below 50 N of vertical force omitted. Unpublished data, Musculoskeletal Biomechanics Research Laboratory, University of Southern California.

weight acceptance) is much higher than that of stair ascent (190% body-weight) (figure 7(a)). This greater value reflects the greater downward acceleration of the body as it is lowered to the next step. On the other hand, the second peak that occurs prior to toe-off is much lower than that of stair ascent (95% body-weight).

The anterior-posterior shear forces demonstrate the same biphasic pattern evident in level walking and ascending stairs; however, the absolute values tend to be somewhat greater (figure 7(b)). The peak anterior shear force acting on the floor during weight acceptance is approximately 15% body-weight, which is similar to the peak posterior force acting on the floor prior to toe-off. The lateral shear force acting on the floor is present also during stair descent with peak values being somewhat greater than those observed during stair ascent (approximately 10% of body-weight) (figure 7(c)).

Despite the fairly large differences in ground reaction forces during stair ascent and descent, the RCOF is quite similar. For example, the RCOF during weight acceptance remains approximately 0.26, while the RCOF just prior to toe-off reaches a maximum of approximately 0.34 (figure 7(d)). Although these data suggest that a slip also is more likely to occur just prior to toe-off, there does not appear to be any greater risk associated with descending stairs as compared to ascending stairs.

2.3.2. *Kinematics of stair ambulation*

2.3.2.1. *Ascending stairs*: During stair ascent, the greatest differences in joint motion (when compared to level walking) occurs at the knee and hip. As the foot makes contact with the stair, the hip and knee are flexed to approximately 60° and the ankle is in about 10° of dorsiflexion (McFayden and Winter 1988, Powers *et al.* 1997) (figure 8(b), (c)). Elevation of the body is accomplished through hip and knee extension, which peaks during late stance (50% of the gait cycle). Early stance-phase ankle dorsiflexion permits tibial progression and accommodates the increased requirement of knee flexion (figure 8(a)). Increased flexion of the hip (60°) and knee (8°) are required during swing to clear the foot, and to place the limb in the appropriate position in preparation for contact with the next step (McFayden and Winter 1988, Powers *et al.* 1997). Swing phase motion of the ankle is similar to that of level walking.

2.3.2.2. *Descending stairs*: During stair descent, contact with the lower step is made with the ankle in approximately 20° of plantarflexion and the knee and hip slightly flexed (10° and 20°, respectively) (Powers *et al.* 1997) (figure 8). Lowering of the body is primarily accomplished through knee flexion, which peaks at about 80° by the end of stance (McFayden and Winter 1988, Powers *et al.* 1997) (figure 8(b)). Hip flexion (30°) also contributes to lowering of the body (figure 8(c)). Progressive ankle dorsifexion is evident throughout stance, reaching a maximum of 15° by approximately 50% of the gait cycle (figure 8(a)). In anticipation for contact with the next step, progressive hip and knee extension is evident and the ankle plantarflexes during swing (Powers *et al.* 1997).

2.3.3. *Joint moments during stair ambulation*

2.3.3.1. *Ascending stairs*: The muscle moments generated at the lower extremity joints during stair ambulation vary significantly compared to level walking. Elevation of the body is accomplished through large hip and knee extensor

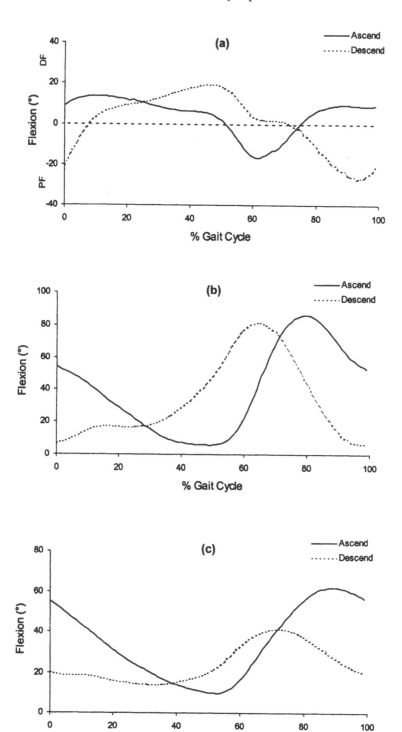

Figure 8. Kinematics for (a) the ankle, (b) the knee, and (c) the hip during stair ascent and descent. Ensemble averaged curves obtained from healthy adults (n = 19). Data taken from Powers *et al.* (1997). DF = dorsiflexion, PF = plantarflexion.

moments, which peak during early stance (approximately 1.0 Nm kg^{-1} at both joints) (Andriacchi *et al.* 1980, McFayden and Winter 1988, Salsich *et al.* 2001) (figure 9(b), (c)). During stair ascent, the ankle demonstrates a plantarflexor moment throughout stance peaking just prior to swing (1.5 Nm kg^{-1}) (Andriacchi *et al.* 1980, McFayden and Winter 1988) (figure 9(a)). The plantarflexor moment controls tibial rotation and provides for push-off during stance.

2.3.3.2. *Descending stairs*:   During stair descent, the lowering of body-weight is accomplished primarily through a knee's strong extensor moment (Andriacchi *et al.* 1980, McFayden and Winter 1988, Salsich *et al.* 2001). The knee moment demonstrates two peaks, one during weight acceptance (early stance) and a larger extensor moment (1.0 Nm kg^{-1}) as the body is being lowered in late stance. The ankle contributes significantly to stability during stair descent as evidenced by a large plantar flexor moment that peaks early during stance (1.4 Nm kg^{-1}) (Andriacchi *et al.* 1980, McFayden and Winter 1988). The hip demonstrates a relatively small extensor moment throughout stance peaking at 0.2 Nm kg^{-1} during weight acceptance (Andriacchi *et al.* 1980, McFayden and Winter 1988).

2.4. *Load carrying using the standard industrial symmetrical 2-handed posture*
Many workers are required to carry loads as part of their daily occupational tasks. Over the years, load lifting and holding tasks have been the focus of research directed towards preventing load handling-related musculoskeletal injuries. For load carrying, researchers have concentrated mostly on muscle activity patterns and physiological strain parameters and less on gait biomechanics and stability parameters. Cham and Redfern (2001b) investigated the effect of carrying relatively light loads (no load, 2.3 and 6.8 kg) on slips- and falls-related gait biomechanics during normal locomotion. In this investigation, load carrying was associated with small but significant decreases in the required frictional properties for safe walking, a finding that was previously reported by Love and Bloswick (1988) for level walking. Kinematic changes associated with load carrying reported by Cham and Redfern (2001b) included minor postural adaptations such as increased knee flexion and slower heel contact velocity along the floor surface.

Myung and Smith (1997) have examined load-carrying effects on specific parameters such as step length and heel velocity during level walking. The authors reported a significant decrease of stride length with increasing load levels, a result that was not confirmed by Cham and Redfern (2001b). This apparent disagreement could be due to the different load levels considered in the two studies. In Myung and Smith (1997) the load level ranged from the no-load condition to 40% of body weight, a far greater load level than the ones investigated by Cham and Redfern (2002b). Myung and Smith (1997) have also concluded that heel velocity at heel contact was not affected by load carrying levels on dry floors, a finding that apparently contradicts Cham and Redfern's results. Differences in the two experimental and analysis procedures could be responsible for this apparent contradiction in the results: (1) Myung and Smith (1997) had a fixed walking speed while Cham and Redfern (2001b) had a natural pace, and (2) Myung and Smith (1997) used the resultant vector of heel velocity, while Cham and Redfern (2001b) investigated the individual components (anteroposterior and lateral) along the floor surface.

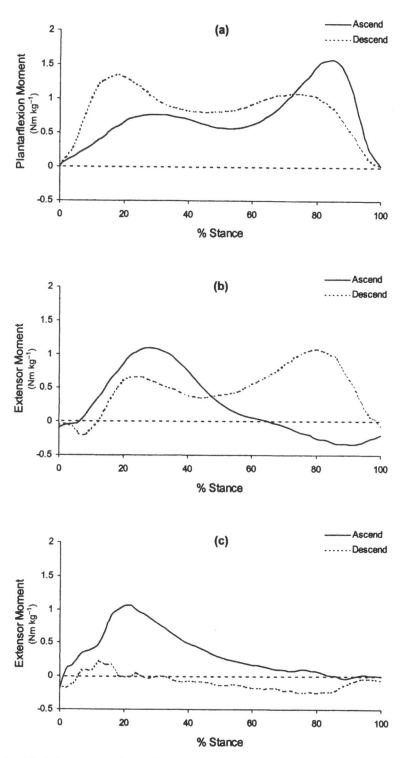

Figure 9. Net joint moments for (a) the ankle, (b) the knee, and (c) the hip during stair ascent and descent. Ensemble averaged curves obtained from healthy adults (n = 10). Data taken from Salsich *et al.* (2001).

3. Biomechanics during slipping

3.1. *Definition of slip from a biomechanical perspective: microslips/macroslips and falls*

During normal gait on dry, non-slippery surfaces, heel sliding along the floor surface has been observed at and shortly after heel contact before coming quickly to a complete stop. This heel motion characterised as 'normal' (Perkins 1978, Cham and Redfern 2002b) or termed 'grip' (Strandberg and Lanshammar 1981) or 'microslip' (Perkins 1978, Leamon and Son 1989), is not detected by subjects. Based on the distribution of slip distances on dry surfaces or the human perception of slipping (Leamon and Li 1990), researchers have used cut-off values of 1 cm (Perkins 1978, Cham and Redfern 2002b) or 3 cm (Leamon and Li 1990) above which the outcome of a contaminated trial was classified as a full slip or 'macroslip' (table 3). Standberg and Lanshammar (1981) used a somewhat more detailed categorisation of contaminated surface trials that did not develop into falls (slip-stick). The so-called slip-stick trials were further divided into three groups: mini-slip (subjects did not detect the slipping motion), midi-slip (slip-recovery trials without 'major gait disturbances') and maxi-slip (slip-recovery with large corrective responses or 'near-fall' trials). As expected, slip distance and peak forward sliding velocity increase with the severity of the slip (table 3). Redfern and colleagues (Hanson *et al.* 1999, Cham and Redfern 2002b) have categorised trials as falls in two cases: (1) the heel kinematic data showed that the heel did not come to a stop after heel contact, and/or (2) the subject lost balance and eventually fell into the safety harness.

3.2. *Kinematics of walking on slippery surfaces*

Biomechanical human reactions to slippery surfaces partially determine the outcome of slipping perturbations (no slip, slip-recovery, slip-fall), and are therefore important to monitor for understanding the complex relationship between gait biomechanics and actual slips and falls incidence. The descriptions of typical slip-recovery events appear consistent across studies (Perkins 1978, Strandberg 1983, Cham and Redfern 2002b), although the magnitude and timing of gait parameters are not always given. Typically, trials leading to a slip event are characterised by higher linear impact heel velocities (not always consistent across Strandberg's subjects), slower foot angular velocities at heel contact and faster sliding heel movements after heel contact, when compared to dry or grip trials (table 3). Generally, subjects are able to slow down the heel to very low velocity levels, often even sliding in the rearward direction. Cham and Redfern (2002b) reported that subjects were always able to rotate their foot down onto the floor and reach foot-flat position regardless of the trial's outcome. As pointed out by Perkins (1978), Strandberg (1983), and Cham and Redfern (2002b), the forward slip starts slightly after heel contact (about 50–100 ms) (for example, see figures 10 and 11). Strandberg (1983) suggested that a slip is likely to result in a fall if the slip distance is in excess of 10 cm or the peak sliding velocity is higher than 0.5 m s^{-1}. Cham and Redfern (2002zb) reported heel velocity of slip-fall outcomes reaching a local maximum (table 3) before subjects attempted to control the slipping motion thus slowing the heel's sliding motion, sometimes even reversing it (as shown in figure 11) to a local minimum of heel. At that time, the heel accelerates again and eventually leads to a fall. This attempt to recover has not been reported by Perkins (1978), but is evident in data presented by Strandberg (1983), although not discussed.

Table 3. Definition and characteristics of trial outcomes for level walking.

Slip distance (SlipDist) (cm) Peak forward velocity of slip (MaxVel) (m s⁻¹)	Perkins[†]	Strandberg[‡]		Leamon and Li[§]	Grönqvist[¶]	Cham and Redfern[††]	
	SlipDist (cm)	SlipDist (cm)	MaxVel (m/s)	SlipDist (cm)	SlipDist (cm)	SlipDist (cm)	MaxVel (m/s)
Normal-slip, microslip or grips on dry surfaces	<1.0	NA	NA	<3.0	NA	<1.0	0.10 (SE 0.01)
Macroslip, slip, slip-recovery, slip-stick	Few cm	Mini-slip 1.2 (0.4) Midi-slip 5.1 (4.7) Maxi-slip >8.6 (3.7)	0.23 (0.04) 0.49 (0.22) 0.56 (0.50)		1.6 (0.4) during 60 ms after heel contact 2.7 (4.7) 60–200 ms after heel contact*	4.0 (SE 3.0)	0.31 (SE 0.06)
Fall, skid-fall	>10–15	NA	> gait speed (1–2 m s⁻¹)	NA	NA	No recovery when Slip-Dist >10.0	0.78 (SE 0.16)**

[†]Perkins (1978), [‡]Strandberg (1983), [§]Leamon and Li (1990), [¶]Grönqvist (1999), [††]Cham snd Redfern (2002b).

*One slip (out of 12 trials) was a likely fall since the total slip distance was 19 cm and the peak slip velocity was 2.1 m s⁻¹ (160 ms after heel contact). All other slip distances were less than 6 cm (range 1.5 to 5.9 cm) and peak slip velocities were less than 0.55 m s⁻¹ (range 0.22 to 0.54 m s⁻¹).

**For fall trials, MaxVel derived before subject attempts a recovery, at which time the heel velocity is brought to a minimum before increasing again, leading to a fall.

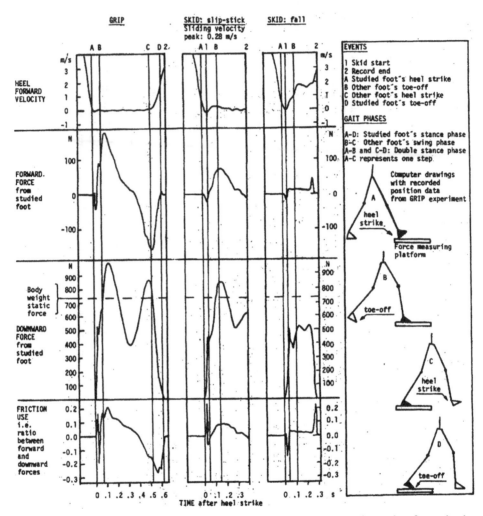

Figure 10.   Typical characteristic profiles of heel velocity and ground reaction forces in the direction of motion reported by Strandberg (1983).

3.3. *Ground reaction forces during slips*

GRF profiles appear to be more varied than kinematic variables across slip trials. However, general characteristics can be identified. First, both peak shear and normal GRFs are reduced during slip events (Strandberg 1983). Second, the transfer of body weight to the supporting leg does not seem to be completed in fall trials. This is evident not only in the shape of the normal forces (Strandberg 1983), but also in the progression of the centre of pressure, which stayed close to the ankle in fall cases (Cham and Redfern 2001b). Third, after a slip has developed, a corrective response or attempt at bringing the foot back near the body, can sometimes be identified as associated with a decrease in the shear forces (25–45% into stance) (Cham and Redfern 2001b).

As mentioned in section 2.1.1, on dry surfaces, the shear to normal force ratio or RCOF has been interpreted as the frictional requirements for a no-slip outcome

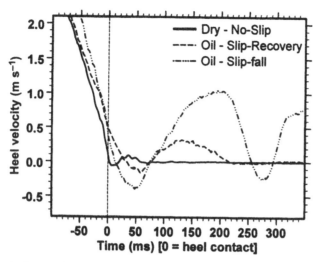

Figure 11. Typical examples of linear heel velocity profile recorded along the direction of motion during dry (no-slip) and oily (slip-recovery and slip-fall) conditions (from Cham and Redfern 2002b).

(Perkins 1978, Strandberg 1983, Redfern and DiPasquale 1997). During slippery trials, the shear to normal ratio was defined as the instantaneous utilised friction (Strandberg 1983) or achievable friction (Hanson *et al.* 1999), which decreased with the severity of the slip (Strandberg 1983, Grönqvist *et al.* 1993). For soapy conditions and level surfaces, for example, Strandberg (1983) has reported achievable COF (ACOFs) as low as 0.02 (a fall case) up to 0.15 (mini-stick), compared to the peak RCOF value of 0.17 for grip trials. Grönqvist found that the time-averaged (100–150 ms after heel contact) ACOF during slippery trials (for seven male subjects) on a level surface was 0.11 ($\pm$0.04) for the 'slip-recovery' condition and 0.04 ($\pm$0.02) for the 'slip-fall' condition (Grönqvist *et al.* 1993).

4. Postural control

4.1. *Joint moment response during slips*

Dynamic analyses have been used in many biomechanical studies on a number of activities such as lifting or carrying to determine the net moments at various joints. However, the joint moments of force of the lower extremity have been most widely researched in gait and balance studies. Moment patterns vary at the hip and knee during walking as a result of the balance control of total limb synergy (Winter 1995). The large changes in hip and knee moment patterns seen during gait on level and inclined surfaces (figure 5), serve not only to generate the power necessary for the task, but also must keep a proper inter-segmental relationship to maintain balance. If the moments are not distributed and co-ordinated among the joints properly, balance would be compromised. Joint moments generated during a slip reflect an attempt of the person to bring the body back into equilibrium.

Joint moments in response to slipping represent the biomechanical reactions to maintain or recover balance. A steady gait pattern will be interrupted at the onset of slip and a rapid balance recovery attempted. This recovery attempt, sometimes termed a protective stepping strategy, often included large moment deviations from the steady gait pattern since the occurrence of such stepping is often unexpected and unrehearsed. Cham and Redfern (2001b) have investigated corrective strategies adopted in an attempt to avoid a fall after a slipping perturbation on oily level surfaces. Increased flexion moment at the knee was identified as the dominant response to slips between 25 and 45% into stance (figure 12(b)). Coincidently, the moment generated at the hip reflected a bias towards extensor activity (figure 12(c)). The ankle joint, on the other hand, acted as a passive joint during fall trials (figure 12(a)). This is due to the centre of pressure's proximity to the heel throughout stance in the fall cases, indicating, as mentioned previously, an uncompleted body weight transfer to the leading foot. Cham and Redfern (2001b) also reported that the corrective movements produced by those moments included increased knee flexion reaction, allowing subjects to rotate the shank forward, restore the ankle angle profile in an attempt to bring the foot back near the body, an effect that was evident in the deceleration of the sliding heel even in the case of slips resulting in falls (figure 11).

4.2. *Postural strategies*

Even though upright posture is inherently unstable and once a slip occurs a fall may appear to be unavoidable even among young adults (Pai 1999), humans still have a wide range of biomechanical responses available to protect themselves from actually falling to the ground. These responses can be both volitional involving conscious efforts and/or automatic involving reflexive reactions. They can be proactive as well as reactive movement strategies that can be implemented prior to, during, or after the loss of balance are experienced (Patla 1993, Woollacott and Tang 1997). Proactive control mechanisms are those that take place before the body encounters a potential threat to stability. A good example is the early detection and avoidance of potentially hazardous situations prior to actual contact (Woollacott and Tang 1997, Tang *et al.* 1999). Another example is the significant reduction in the peak RCOF recorded by Cham and Redfern (2002a) during trials when subjects anticipated the possibility of slips. The RCOF under these conditions were reduced by 16–33%. In addition, the perception of the danger of slipping affected the loading rate on the supporting foot, the joint moments and foot-floor angle at heel contact (Cham and Redfern 2002a).

Most slip and fall incidences occur unexpectedly. After the onset of a slip, a wide range of protective responses can involve both upper and lower extremities, such as grasping, arm swing, hip and ankle motion (ankle/hip strategy) (Horak and Nashner 1986, Horak 1992), and compensatory stepping (Maki and McIlroy 1997), as well as trunk motion. Grasping can be a quite effective recovery response, but one obvious limitation of the grasping strategy is the potential lack of any 'graspable' fixtures where the fall occurs. Even though the correction generated by ankle/hip movement can be produced in standing (Gielo-Perczak *et al.* 1999) or similarly during gait (Woollacott *et al.* 1999), it is often insufficient for protection against a fall. Larger disturbances in standing balance can seldom be restored without the subject's taking a step (Maki and McIlroy 1997). Thus the stepping response has a unique and irreplaceable importance in fall prevention.

The ability to recover is most likely determined by multiple factors in an interactive relationship. It is unclear, however, what factors determine the success rate of recovery in the protective stepping response after onset of a slip, what are the 'tradeoffs' between these factors, and furthermore what are the threshold values (or a

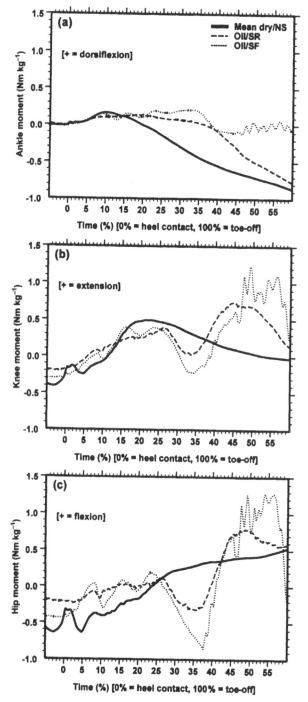

Figure 12. Mean profile of muscle moments generated at the lower extremity joints during stance phase on dry floors compared to typical profiles recorded during slip-recovery (SR) and slip-fall (SF) events on oily floors: (a) ankle moment, (b) knee moment, and (c) hip moment. The ankle moment decreased with the severity of the slip. Knee flexor and hip extensor moments were responsible for corrective reactions attempted between 25 and 45% into stance during slip events (from Cham and Redfern 2001b).

M. S. Redfern et al.

range of threshold values) beyond which a fall is mostly unrecoverable. These factors will probably include those affecting the relationship between the COM and base-of-support (BOS), such as the motion state (distance and velocity) of the slipping foot (Strandberg and Lanshammar 1981, Brady *et al.* 2000, Pavol *et al.* 2000) (figure 13) as well as the step length of the recovery limb (Hsiao and Robinovitch 1999).

4.3. *Joint stiffness control*

Studies of the dynamic behaviour of responses to perturbation while standing have been used to understand the complex responses of persons to slips while walking. Numerous studies by Winter (Winter 1990, Winter *et al.* 1998) have suggested that a body behaves like an inverted pendulum during the initiation of gait or perturbation and the centre of mass (COM) of the body is regulated through movement of the centre of pressure under the feet. They have postulated that the centre of pressure is controlled by ankle plantarflexor/dorsiflexor moment in the sagittal plane and hip abductor/adductor moment in the frontal plane. Winter *et al.* (1998) proposed a relatively simple control scheme for the regulation of upright posture that provides an almost instantaneous corrective response and reduces the operating demands on

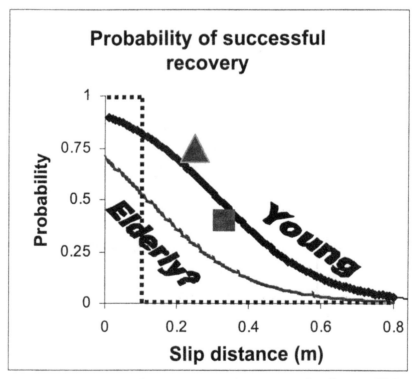

Figure 13.   Strandberg and Lanshammar (1981) suggested a slipping distance of 0.1 m to be the likely threshold for a fall (dotted line). Recent work has shown much higher threshold values (thick line based on Brady *et al.* 2000), where recovery rate reduces from approximately 75% at 0.2 m to just over 10% at 0.6 m. These results are very similar to the observation made elsewhere (square and triangle, Pavol *et al.* 2000). The differences may result from the discrepancies in the methodology. It may be further hypothesized that the recovery rate will be further reduced among the older adults (thin line).

the central nervous system (CNS). Using these assumptions, Gielo-Perczak *et al.* (1999) proposed a mechanical structure of body response, which illustrated the combined effects of stiffness and damping of a subject on the strategy of the control of upright posture. Results showed that three types of postural strategies were performed during quiet standing in a frontal plane at: (1) the ankle joint; (2) the hip joint; (3) a combined strategy using both the hip and ankle joint together. In addition, these postural responses were found to depend on the type of perturbation and joint stiffness. It was observed that the nature of the perturbation must be known by the nervous system before joint stiffness was established. By adjusting joint stiffness, the resonant frequency is shifted to reduce potential resonance from the perturbation. Gielo-Perczak *et al.* (1999) concluded that joint stiffness control is used to help to maintain balance in response to slipping perturbations during walking on slippery surfaces.

5. Conclusions and future research

The biomechanics of slips and falls are an important component in the prevention of injury. This information can be used to develop slip resistance testing methodologies to reflect the frictional properties actually encountered in locomotion. In addition, biomechanical investigations of slips and falls can isolate specific events and times that are critical in determining the differences between slips leading to recovery and those leading to falls. One of the most critical biomechanical factors in slips and falls is thought to be the development of foot forces as the foot comes in contact with the ground. These forces (particularly the shear forces) must be counteracted by the properties of the shoe/floor interface. The ratio of the shear to normal foot forces generated during gait (known as the 'friction used' or RCOF) has been one biomechanical variable most closely associated with the measured frictional properties of the shoe/floor interface (usually the coefficient of friction, COF). Comparing this aspect of gait with measured shoe-floor properties appears to hold great promise for understanding the relationship between walking and potentials for slips and falls. However, there are other biomechanical factors in walking and slipping that also play an important role, such as the kinematics of the foot at heel contact. Slips of the heel naturally occur during most steps, with slip lengths of less than 1.0 cm. These slips (termed micro-slips) occur without the knowledge of the walker. This slipping action of the heel becomes correlated with actual slips noticed by the walker and falls as the slipping distance is increased. Linear motion of the foot coupled with rapid rotations at the ankle at about the time of heel contact make the actual dynamics and trajectories of the heel during these slips complicated, with motions occurring in the forward and rearward directions. The motions and forces at the foot are also variable, depending on the mental set of the walker. If there is a perceived danger of slipping, foot forces and kinematics will change (even if subjects are instructed not to do so). Thus, the biomechanics of walking are subject to the perceptions of the environment by the individual (Grönqvist *et al.* 2001b).

Future research on the biomechanics of slips will be needed to assist in reducing slip and fall injuries. Clearly, one area of future research is to expand our understanding of the shoe/floor contact interactions during slipping (Chang *et al.* 2001a, b). These data can then be used to develop a more 'biofidelic' slip resistance testing device that can measure friction under biomechanically relevant conditions. It is believed that testing slip resistance at the velocities, force levels, pressures and contact times seen in pedestrian walking will greatly increase the predictability of slips by these devices. Thus, knowledge of the kinetics of the foot during walking and slipping is necessary. A second related area of research is to conduct experiments where actual slips occur and relate the biomechanics to slip resistance measurements. While there have been a few studies that have measured the biomechanics of slipping, more research needs to be done. These studies have to carefully control the environments, the instructional set given to the subjects and the gait speed. In addition, data should be transferable to other experiments which are trying to assess the predictability of slip resistance testing devices.

Another future research direction is the investigation of human postural control strategies to prevent falls, including balance reactions when the environment is unknown and when it is known. This might include laboratory experiments investigating stepping responses or moment generations after a slip for different populations, including young adults, older adults, or persons with disabilities. Comparing the capabilities across populations will provide an understanding of the capabilities of people within different environments. The inclusion of research using computer model simulations in concert with the experimental studies may prove to be beneficial as well. The simulations under various conditions can reveal the biomechanical characteristics of the falls that have not been thoroughly demonstrated or understood. For example, one cannot *selectively* alter a subject's muscle strength or functional BOS in order to study the impact of reduced strength or functional BOS on movement stability. Nevertheless, such investigation can be readily performed with the aid of biomechanical model simulation.

A final area of suggested biomechanical research is the investigation of the mechanisms of falls and recovery to guide development of patient-based intervention strategies to prevent fall-related injuries. Such intervention can be achieved by improving both proactive and reactive motor responses. For example, biomechanical studies can be used to suggest methods for improving rehabilitation techniques for the elderly who are at risk for falls.

Acknowledgements

The authors would like to thank John Abeysekera, Patrick Dempsey, Gunvor Gard, Simon Hsiang, Tom Leamon, Raymond McGorry, and Dava Newman for their thoughtful reviews of earlier drafts of the manuscript. Helpful discussions on this subject among participants during the Measurement of Slipperiness symposium were also appreciated. This manuscript was completed in part during Dr. Grönqvist's tenure as a researcher at the Liberty Mutual Research Center for Safety and Health.

References

AMONTONS, M. 1699, De la resistance causée dans les machines. Histoire de L'Académie Royale des Sciences, 266.

ANDERSON, R. and LAGERLÖF, E. 1983, Accident data in the new Swedish information system on occupational injuries, *Ergonomics*, **26**, 33–42.

ANDRIACCHI, T. P., ANDERSSON, G. B. J., FERMIER, R. W., STERN, D. and GALANTE, J. O. 1980, A study of lower-limb mechanics during stair climbing, *Journal of Bone and Joint Surgery*, **62-A**, 749–757.

BRADY, R. A., PAVOL, M. J., OWINGS, T. M. and GRABINER, M. D. 2000, Foot displacement but not velocity predicts the outcome of a slip induced in young subjects while walking, *Journal of Biomechanics*, **33**, 803–808.

BUCZEK, F. L. and BANKS, S. A. 1996, High-resolution force plate analysis of utilised slip resistance in human walking, *Journal of Testing and Evaluation*, **24**, 353–358.

BUCZEK, F. L., CAVANAGH, P. R., KULAKOWSKI, B. T. and PRADHAN, P. 1990, Slip resistance needs of the mobility disabled during level and grade walking, *Slips, Stumbles, and Falls: Pedestrian Footwear and Surfaces*, ASTM STP 1103 Gray B. Everett (ed), (ASTM: Philadelphia, PA), 39–54.

CHAM, R. and REDFERN, M. S. 2001a, Load carrying and gait affecting slip potential, *International Journal of Industrial Ergonomics*. Accepted with revisions.

CHAM, R. and REDFERN, M. S. 2001b, Lower extremity corrective reactions to slip events, *Journal of Biomechanics*. **34(11)**, 1439–1445.

CHAM, R. and REDFERN, M. S. 2002a, Changes in gait biomechanics when anticipating slippery floors, *Gait and Posture*, **15**(2), 159–171.

CHAM, R. and REDFERN, M. S. 2002b, Heel contact dynamics during slip events on level and inclined surfaces, *Safety Science*, **40**(7–8), 559–576.

CHANG, W.-R., KIM, I.-J., MANNING, D. and BUNTERNGCHIT, Y. 2001a, The role of surface roughness in the measurement of slipperiness, *Ergonomics*, **44**, 1200–1216.

CHANG, W.-R., GRÖNQVIST, R., LECLERCQ, S., MYUNG, R., MAKKONEN, L., STRANDBERG, L., BRUNGRABER, R., MATTKE, U. and THORPE, S. 2001b, The role of friction in the measurement of slipperiness, Part 1: Friction mechanisms and definition of test conditions, *Ergonomics*, **44**, 1217–1232.

COULOMB, C. A. 1781, Theory of simple machines. *Memoire de L'Académie Royale des Sciences*, **10**, 161.

COURTNEY, T. K., SOROCK, G. S., MANNING, D. P., COLLINS, J. W. and HOLBEIN-JENNY, M. A. 2001, Occupational slip, trip, and fall-related injuries—can the contribution of slipperiness be isolated?, *Ergonomics*, **44**, 1118–1137.

GIELO-PERCZAK, K., WINTER, D. A. and PATLA, A. E. 1999, Analysis of the combined effects of stiffness and damping of body system on the strategy of the control during quiet standing, *Proceedings of the XVIIth Congress of the International Society of Biomechanics*, Calgary, Canada, 8–13 August.

GRIEVE, D. W. 1983, Slipping due to manual exertion, *Ergonomics*, **26**, 61–72.

GRÖNQVIST, R. 1999, Slips and falls, in S. Kumar (ed.), *Biomechanics is Ergonomics* (London: Taylor & Francis), 351–375.

GRÖNQVIST, R., HIRVONEN, M. and TUUSA, A. 1993, Slipperiness of the shoe-floor interface: comparison of objective and subjective assessments, *Applied Ergonomics*, **24**, 258–262.

GRÖNQVIST, R., ROINE, J., JÄRVINEN, E. and KORHONEN, E. 1989, An apparatus and a method for determining the slip resistance of shoes and floors by simulation of human foot motions, *Ergonomics*, **32**, 979–995.

GRÖNQVIST, R., CHANG, W. R., COURTNEY, T. K., LEAMON, T. B., REDFERN, M. S. and STRANDBERG, L. 2001a, Measurement of slipperiness: fundamental concepts and definitions, *Ergonomics*, **44**, 1102–1117.

GRÖNQVIST, R., ABEYSEKERA, J., GARD, G., HSIANG, S. M., LEAMON, T. B., NEWMAN, D. J., GIELO-PERCZAK, K., LOCKHART, T. E. and PAI, Y.-C. 2001b, Human-centred approaches in slipperiness measurement, *Ergonomics*, **44**, 1167–1199.

HANSON, J. P., REDFERN, M. S. and MAZUMDAR, M. 1999, Predicting slips and falls considering required and available friction, *Ergonomics*, **42**, 1619–1633.

HASLAM, R. A. and BENTLEY, T. A. 1999, Follow-up investigations of slip, trip and fall accidents among postal delivery workers, *Safety Science*, **32**, 33–47.

HORAK, F. B. 1992, Effects of neurological disorders on postural movement strategies in the elderly, in B. Vellas, M. Toupet, L. Rubenstein, J. L. Albarede and Y. Christen (eds), *Falls, Balance and Gait Disorders in the Elderly* (Paris: Elsevier), 137–151.

HORAK, F. B. and NASHNER, L. M. 1986, Central programming of postural movements: adaptation to altered support-surface configurations, *Journal of Neurophysiology*, **55**, 1369–1381.

HSIAO, E. T. and ROBINOVITCH, S. N. 1999, Biomechanical influences on balance recovery by stepping, *Journal of Biomechanics*, **32**, 1099–1106.

JAMES, D. I. 1980, A broader look at pedestrian friction, *Rubber Chemistry and Technology;* **53**, 512–541.

JAMES, D. I. 1983, Rubbers and plastics in shoes and flooring: the importance of kinetic friction, *Ergonomics*, **26**, 83–99.

LANSHAMMAR, H and STRANDBERG, L. 1981, Horizontal floor reactions and heel movements during the initial stance phase, *Eighth International Congress of Biomechanics*, Nagoya, Japan, July.

LEAMON, T. B. and LI, K. W. 1990, Microslip length and the perception of slipping, paper presented at the 23rd International Congress on Occupational Health, Montreal, Canada, September.

LEAMON, T. B. and MURPHY, P. L. 1995, Occupational slips and falls: more than a trivial problem, *Ergonomics*, **38**, 487–498.

LEAMON, T. B. and SON, D. H. 1989, The natural history of a microslip, in A. Mital (ed.), *Advances in Industrial Ergonomics and Safety I, Proceedings of the Annual International Industrial Ergonomics and Safety Conference* (London: Taylor & Francis), 633–638.

LLOYD, D. G. and STEVENSON, M. G. 1992, Investigation of floor surface profile characteristics that will reduce the incidence of slips and falls, *Mechanical Engineering Transaction Institution of Engineers (Australia)*, **ME17** (2), 99–104.

LOVE, A. C. and BLOSWICK, D. S. 1988, Slips and falls during manual handling activities, *Proceedings of the 21st Annual Meeting of the Human Factors Society of Canada*, Edmonton, Alberta, September, 133–135.

MAKI, B. E. and MCILROY, W. E. 1997, The role of limb movements in maintaining upright stance: the 'change-in-support' strategy, *Physical Therapy*, **77**, 488–507.

MANNING, D. P. and SHANNON, H. S. 1981, Slipping accidents causing low-back pain in a gearbox factory, *Spine*, **6**(1), 70–72.

MARPET, M. I. 1996, On threshold values that separate pedestrian walkways that are slip resistant from those that are not, *Journal of Forensic Sciences*, **41**, 747–755.

MCFAYDEN, B. J. and WINDER, D. A. 1988, An integrated biomechanical analysis of normal stair ascent and descent, *Journal of Biomechanics*, **21**, 733–744.

MCVAY, E. J. and REDFERN, M. S. 1994, Rampway safety: foot forces as a function of rampway angle, *American Industrial Hygiene Association Journal*, **55**, 626-634.

MORACH, B. 1993, Quantifizierung des Ausgleitvorganges beim menschlichen Gang unter besonderer Berücksichtigung der Aufsetzphases des Fusses, Fachbereich Sicherheitstechnik der Bergischen Universität—Gesamthochschule Wuppertal, Wuppertal (in German).

MURRAY, M. P., KORY, R. C., CLARKSON, B. H. and SEPIC, S. B. 1967, Comparison of free and fast speed walking patterns of normal men, *American Journal of Physical Medicine*, **45**(1), 8–24.

MYUNG, R. and SMITH, J. L. 1997, The effect of load carrying and floor contaminants on slip and fall parameters, *Ergonomics*, **40**, 235–246.

PAI, Y.-C. and IQBAL, K. 1999, Simulated movement termination for balance recovery: can movement strategies be sought to maintain stability even in the presence of slipping or forced sliding? *Journal of Biomechanics*, **32**, 779–786.

PATLA, A. E. 1993, Age-related changes in visually guided locomotion over different terrains: major issues, in G. E. H. Stelmach, *Sensorimotor Impairment in the Elderly* (Dordrecht: Kluwer), 231–252.

PAVOL, M., RUNTZ, E. and PAI, Y.-C. 2000, Unpublished data.

PERKINS, P. J. 1978, Measurement of slip between the shoe and ground during walking, in *Walkway Surfaces: Measurement of Slip Resistance*, ASTM STP 649, Philadelphia, PA.

PERKINS, P. J. and WILSON, M. P. 1983, Slip resistance testing of shoes—New developments, *Ergonomics*, **26**, 73–82.

POWERS, C. M., PERRY, J., HSU, A. and HISLOP, H. J. 1997, Are patellofemoral pain and quadriceps femoris muscle torque associated with locomotor function?, *Physical Therapy*, **77**, 1063–1075.

PROCTOR, T. and COLEMAN, V. 1988, Slipping, tripping and falling accidents in Great Britain—present and future, *Journal of Occupational Accidents*, **9**, 269–285.

REDFERN, M. S. and ANDRES, R. O. 1984, The analysis of dynamic pushing and pulling: required coefficients of friction, *Proceedings of an International Conference on Occupational Ergonomics*, May 7–9, Toronto, Ontario, 569–571.

REDFERN, M. S. and BIDANDA, B. 1994, Slip resistance of the shoe-floor interface under biomechanically relevant conditions, *Ergonomics*, **37**, 511–524.

REDFERN, M. S. and DiPASQUALE, J. D. 1997, Biomechanics of descending ramps, *Gait and Posture*, **6**, 119–125.

RHOADES, T. P. and MILLER, J. M. 1988, Measurement and comparison of 'Required' versus 'Available' slip resistance, *Proceedings of the 21st Annual Meeting of the Human Factors Society of Canada*, Edmonton, Alberta, September, 14–16.

RICE, D. P. and MACKENZIE, E. J. 1989, *Cost of Injury in the US: A Report to Congress* (San Francisco, CA: Institute of Health and Aging, University of California/Johns Hopkins University Press).

SALSICH, G. B., BRECHTER, J. H. and POWERS, C. M. 2001, Lower extremity kinetics during stair ambulation in subjects with and without patellofemoral pain, *Clinical Biomechanics*, **16**, 906–912.

STRANDBERG, L. 1983, On accident analysis and slip-resistance measurement, *Ergonomics*, **26**, 11–32.

STRANDBERG, L. and LANSHAMMAR, H. 1981, The dynamics of slipping accidents, *Journal of Occupational Accidents*, **3**, 153–162.

SUN, J., WALTERS, M., SVENSSON, N. and LLOYD, D. 1996, The influence of surface slope on human gait characteristics: a study of urban pedestrians walking on an inclined surface, *Ergonomics*, **39**, 677–692.

TANG, P.-F., WOOLLACOTT, M. H. and CHONG, R. K. Y. 1998, Control of reactive balance adjustments in perturbed human walking: roles of proximal and distal postural muscle activity, *Experimental Brain Research*, **119**, 141–152.

TROUP, D. G., MARTIN, J. W. and LLOYD, D. C. 1981, Back pain in industry, a prospective survey, *Spine*, **6**(1), 61–69.

US DEPARTMENT OF LABOR, BUREAU OF LABOR STATISTICS 1997, *National Census of Fatal Occupational Injuries*, 1996, USDL 97-266 (Washington, DC: US Government Printing Office).

US DEPARTMENT OF LABOR, BUREAU OF LABOR STATISTICS 1998, *Lost-worktime Injuries and Illnesses: Characteristics and Resulting Time Away from Work*, 1996, USDL 98-157 (Washington, DC: US Government Printing Office).

WHITTLE, M. W. 1999, Generation and attenuation of transient impulsive forces beneath the foot: a review; *Gait and Posture*, **10**, 264–275.

WILSON, M. P. 1990, Development of SATRA slip test and tread pattern design guidelines, *Slips, Stumbles, and Falls: Pedestrian Footwear and Surfaces*, ASTM STP 1103, (ASTM: Philadelphia, PA). 113–123.

WINTER, D. A. 1991, *The Biomechanics and Motor Control of Human Gait: Normal, Elderly and Pathological*, 2nd edn (Waterloo, Ontario: University of Waterloo Press).

WINTER, D. A. 1995, ABC (*Anatomy, Biomechanics and Control*) *of Balance During Standing and Walking* (Waterloo, Ontario: Waterloo Biomechanics).

WINTER, D. A., Patla, A. E., PRINCE, F., ISHAC, M. G. and GIELO-PERCZAK, K. 1998, Stiffness control of balance in quiet standing, *Journal of Neurophysiology*, **80**, 1211–1221.

WOOLLACOTT, M. H. and TANG, P.-F. 1997, Balance control during walking in the older adult: research and its implications, *Physical Therapy*, **77**, 646–660.

WOOLLACOTT, M., TANG, P.-F. and LIN, S.-I.. 1999, Dynamic balance control in older adults: does limited response capacity lead to falls?, in G. N. Gantchev, S. Mori and J. Massion (eds), *Motor Control: Today and Tomorrow* (Sofia, Bulgaria: Academic Publishing House 'Prof. M. Drinov'), 293–305.

CHAPTER 4

Human-centred approaches in slipperiness measurement

Raoul Grönqvist†*, John Abeysekera‡, Gunvor Gard§, Simon M. Hsiang¶,
Tom B. Leamon¶¶, Dava J. Newman††, Krystyna Gielo-Perczak¶¶,
Thurmon E. Lockhart‡‡, and Clive Y.-C. Pai§§

†Finnish Institute of Occupational Health, Department of Physics,
Topeliuksenkatu 41 FIN-00250, Helsinki, Finland

‡Division of Industrial Ergonomics, Luleå University of Technology, S-971 87
Luleå, Sweden

§Department of Physical Therapy, Lund University, S-220 05 Lund, Sweden

¶Department of Industrial Engineering, Texas Tech University, Lubbock,
TX 79409-3061, USA

¶¶Liberty Mutual Research Center for Safety and Health, 71 Frankland Road,
Hopkinton, MA 01748, USA

††MIT Department of Aeronautics and Astronautics, Cambridge, MA 02142, USA

‡‡Grado Department of Industrial and Systems Engineering, Virginia
Polytechnic Institute and State University, Blacksburg, VA 24061, USA

§§Department of Physical Therapy, University of Illinois at Chicago, Chicago,
IL 60612-7251, USA

Keywords: Slipperiness measurement; Human factors; Postural and balance
control; Slip recovery; Fall avoidance; Safety criteria; Friction thresholds.

A number of human-centred methodologies—subjective, objective, and combined—
are used for slipperiness measurement. They comprise a variety of approaches from
biomechanically-oriented experiments to psychophysical tests and subjective
evaluations. The objective of this paper is to review some of the research done in
the field, including such topics as awareness and perception of slipperiness, postural
and balance control, rating scales for balance, adaptation to slippery conditions,
measurement of unexpected movements, kinematics of slipping, and protective
movements during falling. The role of human factors in slips and falls will be
discussed. Strengths and weaknesses of human-centred approaches in relation to
mechanical slip test methodologies are considered. Current friction-based criteria and
thresholds for walking without slipping are reviewed for a number of work tasks.
These include activities such as walking on a level or an inclined surface, running,
stopping and jumping, as well as stair ascent and descent, manual exertion (pushing
and pulling, load carrying, lifting) and particular concerns of the elderly and mobility
disabled persons. Some future directions for slipperiness measurement and research
in the field of slips and falls are outlined. Human-centred approaches for slipperiness
measurement do have many applications. First, they are utilised to develop research
hypotheses and models to predict workplace risks caused by slipping. Second, they are
important alternatives to apparatus-based friction measurements and are used to
validate such methodologies. Third, they are used as practical tools for evaluating and
monitoring slip resistance properties of footwear, anti-skid devices and floor surfaces.

---

*Author for correspondence. e-mail: raoul.gronqvist@occuphealth.fi

1. Introduction

Data on how to accurately measure risk exposures to slipping and falling hazards seem to be sparse. One of the underlying reasons is the complex nature of slip and fall avoidance strategies such as their dependence upon anticipation of hazards and adaptations of gait to slippery environments (Strandberg 1985, Llewellyn and Nevola 1992, Cham et al. 2000, Redfern et al. 2001). A number of human-centred approaches for the measurement of slipperiness have been utilised to estimate slipping and falling hazards and risks. These approaches have explored initial events from slip onset to foot slide, as well as subsequent events from loss of balance until falling. The output estimates have comprised, among others, perceived sense of slip and slip distance evaluations, slipperiness ratings, heel velocity measurements, heel and trunk acceleration and postural instability measurements, and falling frequency estimations (Strandberg et al. 1985, Leamon and Son 1989, Grönqvist et al. 1993, Myung et al. 1993, Cohen and Cohen 1994a, b, Hirvonen et al. 1994, Chiou et al. 2000). Other approaches have focused on kinematics (spatial movement of the body) and kinetics (ground reaction forces, utilised and required friction) of slipping and falling (Strandberg and Lanshammar 1981, Morach 1993, Redfern and Rhoades 1996, Hanson et al. 1999, Brady et al. 2000, You et al. 2001) or electromyographic (EMG) activity of compensatory muscle responses during simulated slipping (Tang et al. 1998).

Human-centred measurement methodologies form an important complementary to mechanical friction-based test methods which are discussed elsewhere (Chang et al. 2001a,b). The former add a further dimension—the human factor—to the measurement of slipperiness process, and have been widely used to validate mechanical test methods as well (Strandberg 1985, Jung and Fischer 1993, Grönqvist 1999, Leclercq 1999).

The following human-centred approaches to slipperiness measurement will be addressed (figure 1):

(1) 'subjective' approaches such as rating scales, rankings and paired comparisons of floors and footwear, as well as direct observation of protective responses to slipping;
(2) 'objective' biomechanically-oriented approaches such as measuring ground reaction forces, friction usage, body segment movements, joint angles and moments, slip distances and velocities, centre of mass and centre of pressure trajectories, or electromyography; and
(3) 'combined' approaches that comprise subjective evaluations in combination with objective measurements.

The main objective of the present paper is to review the relevant literature on human-centred approaches for the measurement of slipperiness. The role of human factors in slips and falls is discussed, including topics such as awareness and perception of slipperiness, postural and balance control, and adaptation to slippery environments. An overview of friction-based criteria and thresholds for safe walking without slipping is presented, including criteria for some specific work tasks (pushing and pulling, lifting, load carrying) and some special risk groups (the elderly and mobility disabled).

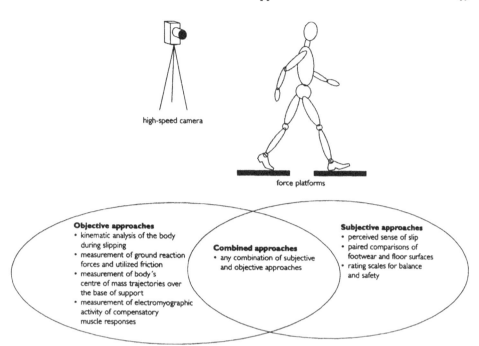

Figure 1. Objective, combined and subjective human-centred approaches for the measurement of slipperiness.

2. The role of human factors in slipping

2.1. *Primary and secondary risk factors*

Injuries due to slips and falls are not purely random events, but rather predictable entities with known risk factors that may be extrinsic (environmental factors), intrinsic (human factors) or mixed (system factors). The primary risk factor for slipping is, by definition (cf. Grönqvist *et al.* 2001), poor grip or low friction between the footwear (foot) and the underfoot surface (floor, pavement, etc.). Static friction is assumed to be important for preventing the initiation of slipping, while dynamic friction would determine whether a foot slide might be recoverable and an injury avoidable, or whether the slip might lead to an injurious fall or any other type of injury, for instance, due to strenuous body movements for regaining balance.

Secondary risk factors ('predisposing factors') for slipping accidents are related to a variety of environmental and human factors, for instance, inadequate lighting, uneven surfaces, incomplete stairway design, non-use of handrails and vehicle exit aids, poor postural control, ageing, dizziness, vestibular disease, diabetes, alcohol intake, and the use of anti-anxiety drugs (Waller 1978, Honkanen 1983, Templer *et al.* 1985, Pyykkö *et al.* 1988, 1990, Sorock 1988, Alexander *et al.* 1992, Malmivaara *et al.* 1993, Nagata 1993, Fothergill *et al.* 1995, Fathallah *et al.* 2000). Merely the slipperiness of the shoe/floor interface may not be a sufficient explanation for falls and other slip-related injuries. The secondary risk factors tend to predispose persons to accidental injuries in slippery conditions and during sudden unexpected changes in slipperiness. The multitude of risk factors and their possible cumulative effects seem to further complicate both slipperiness measurements and the prevention of accidents and injuries due to slipping.

2.2. Postural and balance control

Maintaining postural balance and stability during locomotion is a complex process of position adjustments by muscles and bones acting over joints and controlled by several sensory systems, including vision, vestibular organ, proprioceptive receptors in joints and muscles (e.g. stretch reflexes), and cutaneous receptors with elements such as the pressor receptors of the feet and the velocity- and position-sensitive muscle spindles (Nashner 1983, Johansson and Magnusson 1991). These sensory systems are further controlled by the central nervous systems located in the spinal cord, the brainstem, basal ganglia, cerebellum and cerebral cortex (Horak 1997). Vestibular input governs typically 65% of the body sway during sudden perturbation in standing, while 35% is accounted for by visual, proprioceptive and other input (Pyykkö et al. 1990).

A balance perturbation due to a foot slide constitutes a risk situation, especially if a person's primary task is not maintaining balance. Some common dual-task situations in which locomotion is a secondary task, include walking while reading or talking (on a cellular phone), searching through aisles for a particular object (e.g. grocery shopping), or lifting, lowering and carrying loads while walking. The perseverance of performance in these dual-task situations is critically dependent upon control of the rhythm of locomotion (Danion et al. 1997) and the rhythm of the primary task being attended to (Jones 1976). An accident may occur as a result of a person not being focused on the locomotion, but on perceptual and cognitive processes related to primary task performance. It may also result from a breakdown in rhythm control. High accelerations or decelerations are frequently associated with balance loss, and may cause high inertial forces that have to be supported by the musculoskeletal system.

The way balance is maintained and which balance strategies are sought in a particular situation is dependent on the ability to anticipate postural demands in response to external perturbations, and to move and control the body's centre of mass (COM) back to a position over the base of support (BOS) and the centre of foot support/pressure (COP). A region of stability can be predicted based on the physical constraints of muscle strength, size of BOS, and floor surface contact forces within an environment (Pai and Patton 1997). The alignment of body segments over the BOS must be kept such that the projection of the body COM falls within the boundaries of stability. However, movement termination during a foot slide also depends on the presence of an external braking force that allows the stability region to extend beyond the anterior limit of the BOS in the direction of slipping (Pai and Iqbal 1999). It appears that an increasing friction force might provide such an external braking force, eventually enabling a recovery of stability.

Three strategies for co-ordinating legs and trunk to maintain the body in equilibrium with respect to gravity during standing have been presented (Nashner 1985, Winter 1995): the ankle and hip strategies, and the combined strategy (figure 2). For compensation of antero-posterior sway displacements, the head and the body COM moves in the same direction backwards or forwards during the ankle strategy. This strategy helps to correct small perturbations by using the ankle plantar- and dorsiflexors. During the hip strategy in more perturbed situations or when the ankle muscles cannot act, the body COM moves backwards or forwards opposite to the head movement by means of muscle responses that either flex or extend the hip (Winter 1995). Allum et al. (1998) questioned the major role of lower-leg

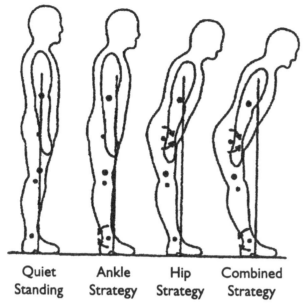

| Quiet Standing | Ankle Strategy | Hip Strategy | Combined Strategy |

Figure 2. Strategies for co-ordinating legs and trunk to maintain body in equilibrium with respect to gravity during standing (adapted from Winter 1995, figure 2.15).

proprioceptive control of posture, especially the dominant ankle strategy, and concluded that postural and gait movements are centrally organised at two levels. The first level involves the generation of a directionally-specific response pattern based primarily on hip or trunk proprioceptive input and secondarily on vestibular input. The second level is involved in the shaping of centrally set multi-sensorial activation patterns, so that movements can adapt to different task conditions.

A sufficient muscle tonus—high enough to maintain postural stability but low enough to facilitate movement—is another prerequisite for the co-ordination of posture and gait. The potential advantage of the stiffness control was recently discussed by Winter *et al.* (1998) during perturbed standing. Their inverted pendulum model assumed that muscles act as springs to cause the COP to move in phase with the COM as the body sways. The COM-COP difference, on the contrary, has been reported to be proportional to the horizontal component of the ground reaction force captured by plantar cutaneous receptors located in the foot sole (Morasso *et al.* 1999). Following latencies in muscle activation after tripping and slipping perturbations have been reported: 120–200 ms for visual control and 60–140 ms for proprioceptive control (Pyykkö *et al.* 1990, Eng *et al.* 1994, Tang *et al.* 1998). However, other studies have claimed that a stiffness control may act almost immediately as the ankle joint angle is changed, causing the COP to move in the same direction as the COM (Winter *et al.* 1998, cf. Redfern *et al.* 2001).

2.3. *Unexpected changes in slipperiness*
The question of whether the risk of slipping injuries is more related to a constantly poor shoe/floor traction or an unexpected, sudden loss of grip due to frictional variation still remains unsolved. The latter case might be the more likely possibility.

Precipitation, temperature and snowfall, and the presence of contaminants or lubricants on the contact surface are important risk factors for occupational and non-occupational slips and falls according to several studies (Honkanen 1982, Merrild and Bak 1983, Lund 1984, Strandberg 1985, Manning et al. 1988, Grönqvist and Roine 1993, Leclercq 1999). Slip-related injuries often seem to occur on wet, dirty, oily, greasy, snowy, icy, or other contaminated walking surfaces. Nevertheless, weak inverse relationships between, for example, seasonal effects (precipitation and cold temperatures) and occupational slips and falls injuries have been reported (Leamon and Murphy 1995).

The role of human factors in the measurement of slipperiness is significant. Strandberg (1985) presented some interesting results from psychophysical walking experiments performed on a continuously slippery triangular path (circumference 12 m). Twelve equally trained healthy male and female subjects were advised to walk as fast as possible over a smooth vinyl (PVC) floor wearing four different shoe types. Other shoe/floor conditions were assessed also, but will not be discussed here. Each test comprised five laps to be walked as fast as possible without slipping and falling. Since cornering was involved in these tests, a higher walking speed required greater friction utilisation. The vinyl floor surface was contaminated and the viscosity of the lubricants was adjusted to either 0.001 N s m^{-2} (water and detergent), 0.01 N s m^{-2} (diluted glycerol) or 0.2 N s m^{-2} (concentrated glycerol). The falling frequencies during five laps were counted and compared to an average time-based friction utilisation (TFU) value during the complete trial and an average force plate-based friction utilisation (FFU) value over one step in the 90° corner of the triangular path (Lanshammar and Strandberg 1985, Strandberg et al. 1985). The results were surprising especially with two of the tested shoe types, Bovinide and Studded (table 1): the test condition with medium TFU (and FFU) caused the highest number of falls (20) while the high and the low TFU (and FFU) conditions caused less falls (6 falls each). The average friction utilisation values showed, as expected, that slipperiness increased as the viscosity of the lubricant increased. However, the medium friction utilisation values in the range 0.20 to 0.25, obtained with the lubricant of medium viscosity (0.01 N s m^{-2}), were surprisingly linked to the highest number of falls during these experiments. (Note: The 'Bovinide' sole was linked to a

Table 1. Viscosity of lubricant, average TFU and FFU, and falling frequency in psychophysical walking experiments on a continuously slippery triangular path covered with a smooth PVC floor (adpated from Strandberg 1985).

	Shoe types				
Parameters	Bovinide & Studded	Bovinide & Studded	Astral & Bovinide & Studded	Bovinide & Studded	Astral & Bovinide & Studded
Viscosity N s m^{-2}	0.001	0.01	0.01	0.2	0.2
Average TFU	0.300	0.205	0.203	0.093	0.078
Average FFU	0.331	0.250	0.248	0.099	0.087
Falling frequency	6	20	26	6	12

Average TFU = average time-based friction utilisation during five laps of the triangular path;

Average FFU = average force plate-based friction utilisation during one step in the 90° corner of the triangular path.

higher falling risk over the 'Studded' sole, which was associated with a greater variation in friction utilisation within steps and over consecutive steps. Thus, the smooth, non-patterned and stiff surface of the 'Bovinide' sole may have reduced the effective contact area in comparison with the flexible and patterned 'Studded' sole.)

One possible explanation for this apparent discrepancy between friction utilisation and frequency of falls might be that the subjects were not able to walk as fast during the slippery, low friction condition (TFU about 0.1) as during the medium friction condition (TFU about 0.2). By adopting a slower 'protective' gait strategy in the low friction condition (i.e. less risk-taking) the subjects were thus able to reduce the number of falls. Then the next question is 'why were the subjects not able to adapt their gait safely during the medium friction condition'? Perhaps because it was a borderline safe/unsafe and deceptive condition, and was hence more difficult to anticipate, which resulted in a greater number of falls. Then why did the high friction condition (TFU about 0.3) cause just as many falls as the low friction condition? This apparent contradiction might be explained by the fact that the subjects were able to walk more quickly over the high friction floor condition, thus challenging their balance more (i.e. taking a higher risk) than during the low friction condition.

2.4. *Postural adjustments and adaptation to slipping risks*

Under constant low-friction conditions, humans typically adopt a protective gait strategy, which involves the combined effect of force and postural changes of the early stance. The subjects take shorter steps and increase their knee flexion, which reduces the vertical acceleration and the forward velocity of the body (Llewellyn and Nevola 1992). As a consequence, the foot impact against the ground should reduce during early stance. Theoretically this strategy should, in the presence of lubricants, minimise the hydrodynamic pressure generation in the lubricant film and the load support between the interacting surfaces (Moore 1972). Hence, a better true contact between the shoe and the underfoot surface should be achieved, which can result in an increased grip and traction (a higher friction coefficient) and a lesser risk for slipping and falling.

A stepping response to a foot slide may have a unique importance in balance recovery and fall prevention (Maki and McIlroy 1997, McIlroy and Maki 1999, Pai and Iqbal 1999). However, it is still unclear what factors might determine the success rate of recovery during a stepping response after the onset of a slip perturbation. These responses can be volitional, involving conscious efforts and/or they can be automatic reflexive reactions. To deal with the risk of falling and injury, the body integrates voluntary movements with so-called 'associated postural adjustments'. These adjustments are involuntary and smoothly organised into the movement repertoire to ensure accurate and harmonious motion. Based on the timing relative to the event of perturbation, the adjustments can be arbitrarily classified into two postural control systems—adaptation and anticipation (Redfern *et al.* 2001)—which in turn may be linked to situation awareness, perception and comprehension of the environment and the ability to project future states. Endsley (1995) gives a thorough definition of situation awareness; one class is concerned with the physical capacity to avoid or accommodate the environmental (ecological) challenges; another class is concerned with the competence to recognise them. Adaptation is reactive in nature and involves the co-ordination of the neuromusculoskeletal system, while anticipation is proactive and entails navigating through complex and often cluttered environments by using multiple sensory inputs to assist in the control and adaptation of gait. The level of situation awareness persons can achieve in a particular environment may dictate whether they anticipate state changes potentially leading to loss of balance. A poor situation awareness may be indicative of adaptive postural adjustments due to perturbations.

Following the above discussion, one might anticipate that it is much easier to adapt one's gait properly when the slippery condition is steady than when rapid and unexpected variation in slipperiness occur. There may not be sufficient injury statistics to confirm this statement, but a study by Merrild and Bak (1983) showed that certain high-risk winter days characterised by drastic temperature changes, precipitation and snowfall, can cause an enormous increase in pedestrian injuries due to falls initiated by slipping. Merrild and Bak (1983) also reported that the proportion of fractures especially in the lower extremities, increased during the high-risk days, and led to a redistribution of injuries in the lower extremities towards their proximal ends. Fractures of the femoral neck and pelvis became more frequent and sprained ankles less frequent than during normal winter days. A recent fall-study made in Norway (Bulajic-Kopiar 2000) confirmed that season affects the incidence of all types of fractures in elderly people, and that slipping on ice or snow seems to be a causal mechanism behind this seasonal effect. In Finland, Honkanen (1982) concluded that the causative role of slippery weather conditions in falls was likely to be more important than any other single factor. He estimated that falls due to slipping during winter months, as compared to summer months, formed 25% 'excess' of all falls on the same level and 47% 'excess' of slip-related falls.

2.5. *Protective movements during falling*

A fall by definition is to descend freely by the force of gravity. A fall occurs when human balance is perturbed beyond a certain recoverable point. When examining forward falls on an outstretched hand from low heights, Chiu and Robinovitch (1998) predicted that fall heights greater than 0.6 m carry significant risk for wrist fractures to the distal radius (the most common type of fracture in the under-75 population). Robinovitch *et al.* (1996) and Hsiao and Robinovitch (1998) studied common protective movements that govern unexpected falls from standing height. They measured body segment movements when young subjects were standing on a mattress and attempted to prevent themselves from falling after the mattress was made to translate abruptly. The subjects were more than twice as likely to fall after anterior translations of the feet (posterior fall) when compared to lateral or posterior translations (anterior falls) of the feet. Since a posterior fall would most likely follow a foot slide during early stance, this study may give us some hints on possible fall mechanisms and protective movements due to slip-induced falls as well. The results by Hsiao and Robinovitch (1998) suggested that body segment movements during falls, rather than being random and unpredictable, involved a repeatable series of responses facilitating a safe landing. Posterior falls involved pelvic impact in more than 90% of their experiments, but only in 23% of the lateral falls and in none of the anterior falls. In the falls that resulted in impact to the pelvis, a complex sequence of upper extremity movements allowed subjects to impact their wrist at nearly the same instant as the pelvis, suggesting a sharing of contact energy between the two body parts. Hsiao and Robinovitch (1999) predicted—using experimental and mathematical models of balance recovery by stepping—that a successful recovery of falling perturbations was governed by a coupling between step length, step execution time, and leg length (cf. Redfern *et al.* 2001).

3. Measurement of slipperiness and falling risks

3.1. *Perception of slipperiness*

During walking one is often unaware of the fact that sliding or creep between the footwear and the floor occurs on contaminated surfaces and even on dry 'non-slippery' surfaces in the very beginning of the heel contact phase (Perkins 1978, Strandberg and Lanshammar 1981, Perkins and Wilson 1983). In fact, Lanshammar and Strandberg (1983) showed that there typically exists an initial spike in the fore-aft component of the ground reaction force immediately after heel strike. The existence of this spike was explained by a small but detectable non-zero backward horizontal motion of the rear edge of the shoe heel. Lanshammar and Strandberg (1983) concluded that since the foot must be strongly decelerated shortly before heel strike an overshoot reaction of the human locomotor system could explain this behaviour.

The lengths of such small slip incidents (also termed micro-slips) on dry, nonslippery surfaces were less than 1 cm, according to Leamon and Son (1989). The tendency of human subjects to such movement on a slippery surface was determined by Leamon and Li (1990), who redefined the term micro-slip to cover a range from zero to 3 cm. Their data indicated that any slip distance less than 3 cm would be detected in only 50% of the occasions, and that a slip distance in excess of 3 cm would be perceived as a slippery condition.

Vision seems to be the only sensory mode that proactively allows a person to identify a slippery floor surface before stepping over it. The other postural control systems may require that one already has walked over a slippery surface for getting the feedback to properly adapt one's gait. Visual control of locomotion has been classified into both avoidance and accommodation strategies (Patla 1991). Avoidance strategies include, for instance, changing the foot placement, increasing ground clearance, changing the direction of gait, and controlling the velocity of the swing foot. Redfern and Schuman (1994) emphasised that temporal control is as critical as spatial control in placement of the foot to maintain balance during gait. Accommodation strategies involve longer term modifications, such as reducing step length on a slippery surface.

3.2. *Biomechanically-oriented approaches*

Ground reaction forces and sagittal plane body movements have been investigated in slippery conditions (as well as in non-slippery baseline conditions) during walking on level and inclined surfaces as well as during load carrying (Perkins 1978, Strandberg and Lanshammar 1981, Morach 1993, Hirvonen *et al.* 1994, McVay and Redfern 1994, Redfern and Rhoades 1996, Myung and Smith 1997, Redfern and DiPasqale 1997). In these studies, researchers especially focused on measuring *leg and heel kinematics*, *normal and shear forces* in the shoe/floor contact surface, and *friction usage* (or required friction) as well as joint angles for the ankle, knee and hip. The effects of *ramp angle* to *forces* at the shoe/floor interface were measured while walking up and down ramps (McVay and Redfern 1994, Redfern and DiPasqale 1997). Also *protective responses* and hand/arm or trunk movements for restoring balance after slipping, *falling frequencies*, and/or *slip/fall probabilities* were examined (Strandberg and Lanshammar 1981, Strandberg 1985, Hanson *et al.* 1999). Some investigators focused on measuring *slip distances* and *micro-slipping* during walking as indicators for slipping and falling hazards (Leamon and Son 1989, Leamon and Li 1990).

Strandberg and Lanshammar (1981) simulated unexpected heel slips when approaching a force platform which was lubricated with water and detergent in 76 trials (61%) out of 124. The trials were categorised into two main groups, grips (85 trials) and skids (39 trials). The skids were split into two categories, slip-sticks (16 trials) and falls (23 trials), while the slip-sticks were finally differentiated into mini-, midi- and maxi-slips. The subjects were unaware of the sliding motion in the mini-slips, in the midi-slips no apparent gait disturbances were observed, but in the maxi-slips compensatory swing-leg and arm motions occurred. The peak sliding velocity was above walking speed $(1-2 \text{ m s}^{-1})$ in the skids that resulted in a fall, but did not normally exceed 0.5 m s^{-1} in the remaining skids called slip-sticks, where the subjects were able to regain balance. These slipping experiments indicated that a slip was likely to result in a fall if the sliding exceeded 0.1 m in distance or 0.5 m s^{-1} in velocity. A recent study by Brady *et al.* (2000) suggested that foot displacement rather than the velocity of the slipping foot would predict the outcome of a slip, and that the threshold values for fall avoidance may be higher than previously thought. Roughly 75% of the subjects in these bare foot slipping experiments over an oily vinyl surface were able to recover balance, when the slip distance was 0.2 m and the slipping foot velocity was 1.1 m s^{-1} (cf. Redfern *et al.* 2001).

Morach (1993) performed slipping experiments on contaminated floors (oil, glycerol, and water) and found that the *horizontal foot velocity* in forward direction immediately (i.e. during 10 ms prior to heel contact) varied between 0.3 and 2.75 m s^{-1} (the average walking speed was 1.5 m s^{-1}) depending on the type of slip, i.e. slip start after a short (more than 26 ms) static position (106 trials), immediate (during less than 6 ms) slip start (300 trials), and unclear (6 to 26 ms) slip start (112 trials). The highest foot velocities occurred on a steel floor with oil as lubricant, when there was an immediate slip start after heel landing.

Winter (1991) and Lockhart (1997) reported a higher heel contact velocity in the horizontal direction for *older subjects* compared with *younger subjects* on dry floor surfaces, even though the walking velocity of the older subjects was slower. On a slippery floor surface (oily vinyl tile), a higher heel contact velocity (figure 3) coupled with a slower transition of whole body centre-of-mass velocity of older individuals significantly affected sliding heel velocity and dynamic friction demand. Consequently, the result was longer slip distances and increased falling frequencies for the older compared to younger individuals (Lockhart *et al.* 2000a, b).

Lockhart *et al.* (2002) conducted a laboratory study to determine how *sensory changes* in elderly people affected subjective assessments of floor slipperiness and how these were associated with friction demand characteristics and slip distance. The results indicated that sensory changes in the elderly increased the likelihood of slips and falls more than in their younger counterparts due to incorrect perceptions of floor slipperiness and uncompensated slip parameters, such as slip distance and adjusted friction utilisation (cf. section 5.4).

Redfern and Rhoades (1996) reported experimental results concerning heel dynamics of subjects during *load carrying* (boxes of varying weights up to 13.5 kg) at three different walking cadences (70, 90 and 100 steps per minute). The surface condition studied was probably dry, but some micro-slipping occurred during the experiments after heel contact. The horizontal (forward) heel velocity decreased from

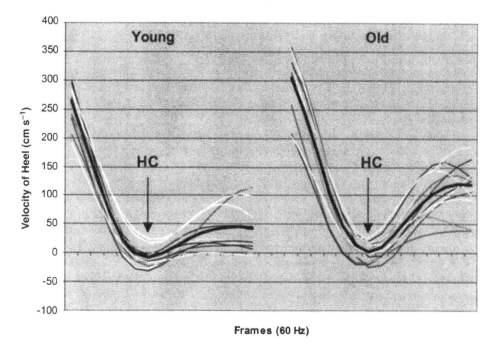

Figure 3. Composite pattern of young and older individual's heel velocity 117 ms before heel contact (HV = heel velocity) and 117 ms after heel contact (SHV = sliding heel velocity) on an oily vinyl floor surface; heel contact (HC) was defined as the time when the vertical ground reaction force exceeded 10 N; the darker line expresses the average pattern of the heel velocities (Lockhart 1997).

a pre-heel contact maximum of 4.5 m s^{-1} at the end of the swing phase to between 0.14 and 0.24 m s^{-1} at heel contact in the beginning of the stance phase. The heel pitch angle at heel landing was between 20 and 25° and decreased to foot flat within about 100 ms after initial contact. The heel came to a complete stop during micro-slip conditions about 100 ms after the impact. Carrying loads showed, according to Redfern and Rhoades (1996), the same dynamic qualities as normal walking. They concluded that load carrying had only minor effects on the heel movement parameters. Recently, Myung and Smith (1997) argued that this was true only for dry conditions while oily floors significantly affected those parameters. They recorded for oily vinyl and plywood floors horizontal heel landing velocities of at least 0.6 to 1.4 m s^{-1} during load carrying experiments with 10 young male subjects. Myung and Smith (1997) also found that stride length was reduced as floor slipperiness and load carrying levels increased (cf. Redfern *et al.* 2001).

3.3. *Psychophysical and subjective approaches*
Human-centred approaches for the measurement of slipperiness may be psycho-physical in nature. A perceived magnitude of 'slipperiness' can be quantified on a psychophysical scale using 'foot movement' or 'postural instability' as the physical stimulus. The stimulus can be measured subjectively using opinions and preferences but can be measured objectively too. Objective measures are, for instance, video filming or high-speed imaging of gait and may also include ground reaction forces obtained with force platforms. Human-centred approaches may involve simulta-

neous acquisition of objective biomechanical data and subjectively perceived data (Strandberg 1985, Grönqvist et al. 1993). Subjective opinion data is quantitatively treated on an ordinal (category) scale, while a nominal scale can be used for analysing motion and force data (slip distance, slip velocity, friction usage, joint angles, etc.). Engström and Burns (2000) spoke for *psychophysical scaling* (continuous ratio scales) as an alternative to common category scaling of opinions and preferences.

A number of 'purely' subjective approaches (e.g. paired comparisons) have been applied to measure slipperiness. Human subjects seem to be capable of differentiating the slipperiness of *floors* (Yoshioka et al. 1978, 1979, Swensen et al. 1992, Myung et al. 1993, Chiou et al. 1996) and *footwear* (Strandberg et al. 1985, Tisserand 1985, Nagata 1989, Grönqvist et al. 1993) in dry, wet, or contaminated conditions. Cohen and Cohen (1994a, b) pointed out that tactile sliding resistance cues are the most sensitive predictors of the coefficient of friction under various experimental conditions but particularly on wet surfaces. Leamon and Son (1989) and Myung et al. (1992) suggested that measuring micro-slip length or slip distance during slipping incidents might be a better means to estimate slipperiness than the apparatus-based friction measurement techniques. Recently Chiou et al. (2000) reported findings of workers' perceived sense of slip during standing task performance (e.g. a lateral reach task) and further related their sensory slipperiness scale to subjects' postural sway and instability. They found that workers who were cautious in assessing surface slipperiness had less postural instability during task performance.

Skiba et al. (1986), Jung and Schenk (1989, 1990) and Jung and Rütten (1992) evaluated walking test methods used for measuring the slipperiness of floor coverings and safety footwear on an *inclined plane* on dry, wet and oily surfaces in the laboratory. The inclination angle at a point when walking down the ramp became unsafe gave the subjective estimate for slip resistance by transforming it geometrically to a friction value. These papers also discussed the validity and reliability of such tests, use of standard reference materials and separation characteristics for choosing a limited number of test subjects for standardised slipperiness measurements.

Subjective and combined human-centred approaches have been utilised to assess footwear *friction on ice*. At least two test rigs have been developed by Bruce et al. (1986) and Manning et al. (1991): the first test rig consists of a tubular metal frame with four legs, fitted with large castor action wheels. A test subject, standing on both feet on an icy surface, is dragged across the substrate. Bruce et al. (1986) conducted tests at an ice skating rink and the horizontal (frictional) force in the shoe/ice interface was measured by a load cell (spring balance) at a low sliding velocity. The frame of the rig prevented the subject from falling. The second technique, by Manning et al. (1991), can be applied to measure the slipperiness of a footwear/ice interface as well as a footwear/floor interface. In this method, a subject is walking (backward steps) on a surface to be assessed while pulling against a spring and supported by a fall-arrest harness. The subject is also protected by two handles suspended from a pulley that moved freely on an overhead rail. The load cell is positioned between the harness belt of the subject and a rigid base (e.g. a wall). Manning and Jones (1993) modified this walking traction rig for mobile field use as well. Since the resisting force is not measured in the shoe/floor interface, the load cell measures indirectly the maximum frictional force before feet will slip. Scheil and

Windhövel (1994), who criticised the validity and precision of this method, evaluated the latter mobile version of the above method on various floor surfaces attached to a force platform. The walking action was abnormal and the friction readings were biased by inertial forces that increased the load cell reading (horizontal force) compared to the measured friction force in the shoe/floor interface.

3.4. *Rating scales for balance and walking safety*

3.4.1. *Balance and functional abilities*:   Rating scales, for example Bohannon's ordinal scale for standing balance (Bohannon and Leary 1997) and the 'Timed Up and Go' test (Podsiadlo and Richardsson 1991) have been used to assess balance capabilities of the individuals, but no perfect standard exists among these methods. An objective test often used today is the Berg balance test (Berg 1989, Berg *et al.* 1992a). Combined with a test of walking speed the Berg balance test shows a high sensitivity and specificity as a screening method (Berg *et al.* 1992a). Berg's balance scale is an instrument for quantitative evaluation of balance, a scale with 14 moments, testing static and dynamic balance with increasing difficulty. All moments are described in a manual with grades from 0 to 4 with a maximum of 56 points. The test takes 20 min to perform. The Berg balance scale has good concurrent validity with many other methods used in this area; the Bartel Index for activities of daily life, Tinetti's sub-scale for balance (Tinetti *et al.* 1986), the Fugl-Meyer scale for isolated movements and balance (Fugl-Meyer *et al.* 1975), the 'Get-up and Go' test (Mathias *et al.* 1986), the 'Timed Up and Go' test (Podsiadlo and Richardson 1991) and a test of functional mobility (Berg *et al.* 1992b).

The test 'Get-up and Go' measures a person's risk of falling according to a 5-grade scale; normal, very slightly abnormal, mildly abnormal, moderately abnormal and severely abnormal (Mathias *et al.* 1986). A person is observed while raising from a chair with armrests, walking 3 m and then returning to the chair. The test focuses on many basic functional aspects and is easy to perform. The 'Timed Up and Go' (Podsiadlo and Richardsson 1991) is a balance test focusing on walking speed and functional ability. The time to perform the test from leaving the chair until sitting on the chair again is measured by a stop-watch.

A functional perspective is important when dealing with balance problems in a test situation. Basic functional abilities for people of all ages are the ability to go and rise from bed, sit down and rise from a chair or toilet, and to walk a few steps (Isaacs 1985). These abilities are prerequisites for elderly people to live independently in their own home or in open home care without assistance. Functional abilities are important for the evaluation of rehabilitation effects in patient groups. For that purpose, an index of muscle function and a battery of functional performance tests for the lower extremities have been developed. The total index can be divided into four separate areas: pre-tests of general functioning; muscle strength; muscular endurance; and balance and co-ordination (Ekdahl *et al.* 1989).

3.4.2. *Evaluation of icy surfaces and anti-skid devices*:   Methods to describe functional problems in walking on different slippery surfaces during winter conditions have been developed by Gard and Lundborg (2000) as rating scales for evaluating walking safety and balance, and as observation scales to observe posture and movements during walking. The methods were then used to investigate functional problems when wearing different anti-skid devices (attached to shoes) for slip and fall protection. First, rating scales for perceived walking safety and balance were developed and tested for reliability. Inter-reliability tests of these scales were done from video-recordings of walking with different anti-skid devices on a number of surfaces in experiments done by two experienced physical therapists. The percentage of agreement between the physical therapists was 86% (walking safety) and 88% (walking balance). Second, four rating scales for evaluation of observed walking movements were developed by a physical therapist, trained in movement analysis. The dimensions evaluated were: (1) walking posture and movements including normal muscle function in the hip and knee; (2) walking posture and movements in the rest of the body (head, shoulders and arms); (3) heel strike; and (4) toe-off. All four dimensions were evaluated by observation scales ranging from 0 to 3. The inter-reliability of these four observation scales were measured as the percentage of agreement between the physical therapists and was 85, 80, 86 and 85%, respectively (Gard and Lundborg 2000).

Abeysekera and Gao (2001) performed practical walk tests using a 5-point rating scale to evaluate slipping risks on a number of icy and snowy surfaces when wearing different types of footwear. Such walk tests may be used to assess the performance of footwear, or anti-skid devices, but also to study the human responses involved in slipping accidents. Gard and Lundborg (2001) carried out practical tests of 25 different anti-skid devices on the Swedish market, on different icy surfaces with gravel, sand, salt or snow on ice, and with pure ice. The anti-skid devices were described according to each subject's perception of walking safety, walking balance and priority for own use. The posture and movements during walking were analysed by an expert physical therapist. One of the tested anti-skid devices was judged to be good regarding walking safety and balance and was chosen by subjects for their own preferred use (Gard and Lundborg 2001).

3.5. *Measurement of sudden movements*

Slipping may involve rapid movements resulting from a person's effort to regain balance. A method has been developed to detect such movements by measuring trunk acceleration during walking (Hirvonen *et al.* 1994). By attenuating normal movement signals by band-pass filtering, it became possible to discriminate signals caused by unexpected movements. The portable equipment consisted of a small acceleration transducer, a pre-amplifier and a pocket computer (Hirvonen *et al.* 1994). Unexpected trunk movements during slipping of 20 male volunteers who walked at two speeds, normal and race walking, along a horizontal track were monitored. The peak acceleration levels of the trunk increased significantly in slipping incidents compared to normal or race walking without slipping, both in the antero-posterior and medio-lateral directions. The peak accelerations varied from 0.5 to 4.5 g ($1 \, g = 9.81 \, \mathrm{m \, s^{-2}}$) during slipping, while the accelerations were less than 0.5 g during walking without slipping. The mean peak accelerations of the trunk during slipping incidents were of the order 1.3 g for the antero-posterior and 1.0 g for the medio-lateral directions, respectively. The levels were significantly higher than during reference non-slipping conditions. The seriousness of these slip incidents was observed using video filming: no observable slip; controlled slip; vigorous slip; extremely vigorous slip. Most experiments resulted in either controlled or vigorous slips.

Questionnaires can be used to study the risk of accidents and the role of sudden movements at work (Hirvonen *et al.* 1996). Workers were asked to subjectively rate the risk of accidents in their work tasks associated with walking on slippery or untidy surfaces, on uneven surfaces, and on stairs. For each question four alternatives of the risk were given: not at all (score 1); a little (score 2); moderately (score 3); much (score 4). Based on these four questions, a sum score of the risk of accident was calculated and classified into three categories: low (score 4–8); moderate (score 9–12); high (score 13–16). In a follow-up intervention at the workplace, a total of 297 unexpected incidents occurred during which the trunk acceleration level exceeded 1.0 g. The number of alarms (i.e. acceleration levels exceeding 1.0 g) was significantly greater for the high risk compared to the low risk category. However, the intensity of sudden movements, measured as the peak acceleration of the trunk, did not differ between the three self-assessed categories of accident risk.

3.6. *Comparative evaluations of test methods*

At least two studies have reported comparative evaluations of test procedures involving human subjects (Jung and Fischer 1993, deLange and Grönqvist 1997). The objective of the first study was to investigate the validity of mechanical laboratory-based test methods for measuring slip resistance of safety footwear when the same test protocol and procedure was applied in each participating laboratory. The second study aimed at bridging the gap between human-centred test procedures and mechanical slip test methods.

Ten subjective and combined human-centred test procedures were compared in the first study (Jung and Fischer 1993). Male subjects (4 to 8 depending on the test) wore five different types of safety footwear and walked over a smooth stainless steel surface contaminated with viscous glycerol or a mixture of water and wetting agent. The walking tests were performed on a straight level surface (3 paired comparison methods) or an inclined surface (7 walk-test methods down or up a ramp). The paired comparison methods yielded either a subjective scoring, a friction usage value, or a slip and fall frequency as the outcome (footwear rating). The outcomes of the ramp tests were either an average maximum inclination angle for safe walking (6 methods) or a paired comparison scoring using two fixed inclination angles (1 method). The rank correlation coefficients between the footwear ratings obtained with these seven human-centred test methods varied from 0.90 to −0.70. Of all the 44 comparisons only six correlation coefficients (14%) between test methods were statistically significant at the 95% probability level. Only two of these significant results were obtained between different types of walking tests, one level surface test and two similar ramp tests. A major limitation of this inter-laboratory experiment was that the assessed footwear did not exhibit large differences in terms of their slip resistance. The validity of the mechanical slip tests could not be confirmed in this study.

Three subjective and combined test procedures were compared in the second study (deLange and Grönqvist 1997). All three tests were also involved in the first study by Jung and Fischer (1993): an inclined surface (walking down a ramp) applying ramp angle as safety criterion (Jung and Schenk 1990) and two level surface test procedures applying subjective scorings based on paired comparisons of footwear as safety criteria. The level surface test methods used two different approaches. The first method was based on human action while walking, stopping and accelerating on a slippery surface (Tisserand 1985) and the second method was based on the heel landing phase when the subject stepped from a slip-resistant surface onto a slippery surface (Grönqvist *et al.* 1993). Male subjects (2 to 7 depending on the test) were wearing six different types of footwear for professional use and walked over a smooth stainless steel surface or a rough vinyl flooring contaminated with viscous glycerol or a mixture of water and detergent. The water and detergent conditions were assessed only with the ramp test method, so that no comparative data between the methods was available for this condition. The results of the two level surface test methods differed significantly for test condition steel/glycerol, probably due to differences in performing the tasks (walking at a constant pace vs. accelerating, etc.). However, the ratings for two test conditions (steel/glycerol, PVC/glycerol) between the second level surface method and the ramp test method were similar despite the two different approaches (walking on a level vs. inclined surface).

4. Modelling slip recovery and fall avoidance

4.1. Difference between static and kinetic friction

Tisserand (1985) suggested using a simple biomechanical model (considering a mechanical equilibrium of forces at the moment the foot strikes the ground) that the slipping velocity (v) is a function of the difference between the static (F_s) and kinetic (F_k) friction forces:

$$v \approx \frac{1}{M}(F_s - F_k) \cdot t$$

where M is the mass of the body parts in motion and t is the time.

Hence, Tisserand (1985) concluded that for a given coefficient of kinetic friction the seriousness of the fall would be directly proportional to the coefficient of static friction. Tisserand also presented experimental subjective evaluation data in support of his reasoning. In fact, Tisserand went even further in his analysis allowing him to assume that this relationship is valid even if there is no initial static phase as the heel strikes the ground. Tisserand (1985: 1030) completed the analysis with the following statement that 'preventing initial slipping of the foot requires a high or sufficient static friction force, while limiting slip velocity to avoid loss of balance requires a small difference between static and kinetic friction forces and that the latter is always required to prevent falling and injury'.

4.2. Critical slip distance

Strandberg and Lanshammar (1981) estimated that the critical sliding velocity leading to falling after a heel slip was about 0.5 m s^{-1}, and that the required minimum kinetic friction coefficient was about 0.2 during normal level walking. If the above figures are accepted as critical for a hazardous slip and fall, then the boundary slip distance s between an avoidable and an unavoidable fall would be about 6 cm (Grönqvist et al. 1999), since:

$$s = \frac{v^2}{2g\mu}$$

The above equation, where g is the acceleration of gravity, v is the velocity of sliding, and μ is the coefficient of friction, is derived from the work done by the frictional force. This equation, which governs the distance required to bring a moving object to a stop by friction, is based on the assumption of a constant initial sliding velocity. Nevertheless, it indicates that if the coefficient of friction increases for example to 0.4, then the boundary slip distance would be reduced to 3 cm, which would be perceived to be slippery by only 50% of the subjects in a slipping experiment (cf. section 3.1). Obviously, an increasingly safer situation from the point of view of balance recovery would follow. On the contrary, if the critical values ($s = 0.2$ m and $v = 1.1$ m s^{-1}) for slip distance and velocity suggested by Brady *et al.* (2000) would be applied in this equation, then the minimum coefficient of friction in the interface would need to be about 0.3 for a fall recovery.

4.3. *Walking speed and step length*

Gait and anthropometric parameters such as length of stride may, however, affect the above critical sliding velocity of the heel. In fact, Strandberg (1985) presented a falling criterion based on the biomechanical skidding data by Strandberg and Lanshammar (1981), using a simple biomechanical inverted pendulum model of the human body during a single stance phase. The model predicted that the maximum permissible (without falling) sliding velocity (v_s) increases with walking speed (v_w) and decreases with step length (L).

The model's falling criterion is:

$$v_s > kv_w - 0.5cL\sqrt{\frac{Mgh}{J}}$$

where k and c are constants and M is the mass of the inverted pendulum, centred the distance h above the heel/floor contact point, and with J the mass moment of inertia about this contact point.

A higher walking speed (v_w) may be favourable, if the step length (L) is kept constant and if no cornering forces are required. The horizontal velocity of the body COM must be sufficiently greater than the sliding velocity of the BOS. Otherwise, the COM velocity relative to the BOS velocity will become negative before COM has reached a position above the BOS, and a backwards falling motion will begin. This gait pattern with short steps in comparison to the walking speed will also result in a reduction of the body's COM vertical acceleration, whether it is the primary aim or not. The relationship between walking speed, step length and friction demand has been investigated by Lindberg and Stålhandske (1981). Recently, You *et al.* (2001) confirmed that the displacement and velocity of the COM with respect to the BOS can be used to discriminate slip/non-slip incidents of the heel in barefoot walking over a slippery soap patch. During the critical double-support period from heel strike to contra-lateral toe-off, a smaller displacement and a faster velocity of the COM were important for regaining balance.

5. Criteria for safe friction

This section will bring some insight to basic safety criteria, frictional demands, and minimum friction thresholds proposed for a number of activities and work tasks including walking, running, and manual exertion. The criteria for some special risk groups (mobility disabled and the elderly) are also discussed. Frictional demands are related to either a baseline non-slip 'required friction' or a more global 'friction usage' (i.e. 'utilised friction') criterion, which is independent of whether a slip occurs or not (cf. Grönqvist *et al.* 2001a).

5.1. *Level walking*

The frictional demand, based on human experiments during normal level walking, has been found to vary between 0.15 and 0.30 (Perkins 1978, James 1980,

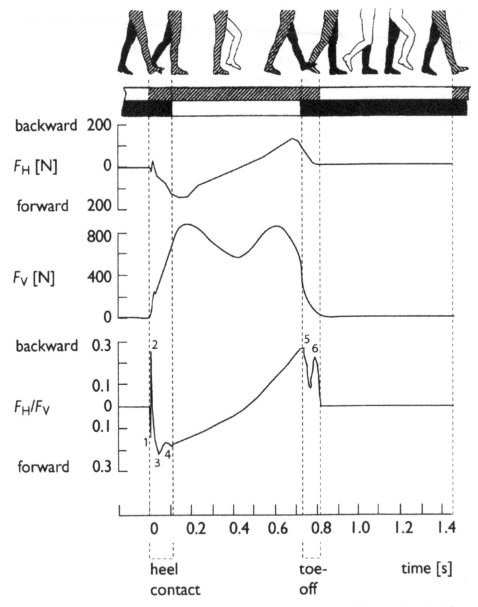

Figure 4. Gait phases in normal level walking with typical horizontal (F_H) and vertical force (F_V) ground reaction components and their ratio, F_H/F_V, for one step (right foot). Critical from the slipping point of view are the heel contact (peaks 3 and 4) and the toe-off (peaks 5 and 6) phases (Grönqvist *et al.* 1989).

Strandberg and Lanshammar 1981, Bring 1982, Skiba *et al.* 1983). See also Redfern *et al.* (2001) who discusses the role of safety criteria based on ground reaction forces during locomotion without slipping and during slips. The significance of the horizontal (shear) to vertical ground reaction force ratio (F_H/F_V) is that it indicates where in the step cycle a slip would most probably start (figure 4). However, if a foot slide starts, the evolution of frictional shear forces in the shoe/floor interface would determine whether the slip can be retarded or stopped and balance recovered. Hence, the measured friction coefficient should always be greater than the utilised or the required friction coefficient (cf. Grönqvist *et al.* 2001a, Redfern *et al.* 2001). Hanson *et al.* (1999) applied the difference between measured and required friction as safety criterion instead of the friction ratio, which was proposed by Carlsöö (1962).

Grönqvist *et al.* (2001b) found that the contact time related variation in utilised friction in the presence of a slippery contaminant was large (possibly due to gait adaptations) in response to the reduced available (measured) friction between the interacting surfaces. In this case, a single safety limit for the friction coefficient, such as the maximum peak value after heel contact, may not be an appropriate discriminator between safe and dangerous conditions. The evolution of the friction coefficient over contact time seems to be at least equally important.

Perkins (1978) reported both 'maximum' and 'average' peak values for the frictional demand. Strandberg and Lanshammar (1981) measured the friction usage peak, F_H/F_V, approximately 0.1 s after heel strike. The peak value in their experiments was on the average 0.17 when there was no skidding (grip), 0.13 when the subject was unaware of the sliding motion or regained balance (slip-stick), and 0.07 when the skid resulted in a fall. Kinetic friction properties appeared to be more important than static ones, because in most of their walking experiments the heel slid upon first contact even without a lubricant. On the other hand, Strandberg *et al.* (1985) and Grönqvist *et al.* (1993) also reported stance time-averaged friction usage values for estimating frictional demands during walking over continuously and unexpectedly slippery surfaces, respectively. The values reported by Strandberg *et al.* (1985) were based on the mean stance time for one step (approximately 0.5–0.7 s). These values were of the order 0.25 for the safest experiments on a smooth steel surface with glycerol as contaminant. In comparison, the mean F_H/F_V ratios during time-interval 100–150 ms after heel contact, for the safest experiments without falling, reported by Grönqvist *et al.* (1993) were much lower (between 0.11 and 0.13) in the same test condition (steel/glycerol).

Strandberg (1983) favoured 0.20 as a safe limit for the coefficient of kinetic friction in level walking, indicating that he added a safety margin to the measured frictional demand (cf. peak ratios of the 'grip' and 'slip-stick' trials above). Nevertheless, he pointed out that the adequate value was depending greatly on anthropometric and gait characteristics as well as the method of measurement. He found that friction properties were most important for preventing falls at sliding velocities below 0.5 m s^{-1}. In contrast, static friction values proposed in the USA in the mid-1970s used 0.40 to 0.50 as lower limits for safe walking (Brungraber 1976). These values are not in line with the actual frictional demands based on human walking on level surfaces, and thus may be more an indication of practical eligibility for the test methods in question.

5.2. Other modes of locomotion

In general, minimum coefficient of friction limit values should be correlated to normal variability of human gait, since walking speed, stride length and anthropometric parameters may greatly affect the frictional demands during locomotion (Carlsöö 1962, James 1983, Andres et al. 1992, Myung et al. 1992). McVay and Redfern (1994) found that the mean across subjects of the peak required friction increased from about 0.25 to 0.50 as ramp angle increased from 0° in level walking to 20° on an *inclined surface*. Their study indicated that geometric predictions based on ramp inclination angles did not fully explain the actual changes in frictional demands and resulted in excessively high values for the required friction limit. Walking up the ramp produced greater frictional demands than walking down the ramp. Ground reaction forces on inclined surfaces are discussed extensively by Redfern et al. (2001).

Depending on the type of movement (walking on a *ramp* or on *stairs*) James (1980) referred to limit values between 0.15 and 0.40. Harper et al. (1967) and Skiba et al. (1983) referred to limit values between 0.30 and 0.60 during *stopping* of motion, *curving* and walking on a *slope*. Later Skiba (1988) defined that the safety limit for the kinetic friction coefficient, based on the forces measured during human walking and the social acceptance of the risk of slipping, would be 0.43 at sliding speeds of at least $0.25 \, \text{m s}^{-1}$. At heel strike in *running* gait the peak F_H/F_V is typically slightly greater (about 0.30) compared to walking, and this difference is even greater at toe-off when the force ratio peak is about 0.45 (Vaughan 1984). *Jumping* during exiting commercial tractors, trailers and trucks can produce large impact forces in the shoe/ground interface (Fathallah and Cotnam 2000) and may increase the friction demand as well as the risk of slipping (Fathallah et al. 2000). The average peak required friction range was from 0.13 to 0.33 depending on the vehicle type, jumping height, and the use of safety aids such as steps and grab-rails.

Skiba et al. (1985) reported that the peak friction usage at foot contact during *stair ascent* was lower than during normal level walking (0.17 versus 0.21). During *stair descent*, the peak friction usage at foot strike was lower than during normal level walking (0.12 versus 0.21) but higher during the push-off phase (0.34 versus 0.20) according to the same study. Christina et al. (2000) found that the peak frictional demand at foot strike during stair descent was remarkably similar to level walking (i.e. between 0.30 and 0.32) but that the required friction was lower during the push-off (0.26). Note that the reported friction usage values were contradictory in these two studies. Redfern et al. (2001) also discusses ground reaction forces during stair ambulation.

5.3. Manual exertion

Kroemer (1974) measured horizontal *push and pull forces* when subjects were standing in working positions on various surfaces. Forces exerted while braced against a footrest or wall were compared against the forces exerted while standing on high, medium and low traction surfaces. The mean push and pull forces exerted were reduced considerably depending on the slipperiness of the surfaces: the forces exerted were roughly 500–750 N for the braced situation, 300 N for the high traction surface (static friction coefficient $\mu > 0.9$), 200 N for the medium traction surface ($\mu \sim 0.6$), and 100 N for the low traction surface ($0.2 < \mu < 0.3$). Ciriello et al. (2001) investigated maximum acceptable horizontal and vertical forces of *dynamic pushing* on high and low friction floors. They found that the 'required' friction coefficient (i.e.

the horizontal to vertical shoe/floor force ratio required to sustain a push-cart movement) was 0.32 for the high friction floor but only 0.19 for the low friction floor. However, push duration on the low friction floor was significantly longer and slip potential greater than on the high friction floor.

Grieve (1979, 1983) examined limitations of performance due to friction and static friction limits for avoiding slipping during manual exertion (*lifting, pushing and pulling*). Grieve found that static manual exertion can create unavoidable slips due to high frictional requirements (even > 1) in some conditions, and concluded that more efforts should be concentrated on the events that follow the foot slide. Zhao *et al.* (1987) showed that the determination of slipping risks associated with *lifting on inclined surfaces* should not be based solely on the slope angle. Dynamic lifting increases the ratio of tangential (shear) to normal forces compared to the static condition. Consequently, Zhao *et al.* (1987) suggested that slope angle must be smaller than the friction angle of the shoe/ground interface. For example, for a slope angle α of 15° the peak force ratio of shear to normal forces would be 0.27 ($= \tan \alpha$) for the static case but in the range of 0.30 to 0.36 for dynamic lifting.

Kinoshita (1985) examined the effects of light and heavy loads on certain gait parameters and found that the shear and normal components of the ground reaction force significantly increased compared to habitual walking. However, the frictional requirements in terms of their ratio did not seem to change due to *different loads* and *carrying systems* (double-pack and back-pack). Analysis of spatial and temporal parameters of the gait patterns revealed that only the double and single support periods of stance were affected by the changes in load. The double support period, expressed as a percentage of the total support period, lengthened and the single support period shortened significantly as the load increased.

5.4. *Mobility disabled and the elderly*

Buczek *et al.* (1990) emphasised that the slip resistance needs for *mobility disabled* may be greater than for able-bodied persons. Their study indicated that the required coefficient of friction near touch-down for the unaffected side of the mobility disabled person was significantly higher (average 0.64) than for the able-bodied (average 0.31) regardless of the speed (slow or fast) of walking, whereas no difference was observed for the push-off phase.

Christina *et al.* (2000) found in *stair descent* that the frictional demand at foot touch-down was lower for the *elderly* (0.27–0.28) than for the younger subjects (0.30–0.32), indicating a safer stair descent strategy chosen by the elderly (cf. section 5.2).

On a dry *level-walking surface* (outdoor carpet), Lockhart *et al.* (2000a), found no statistically significant differences in required coefficient of friction characteristics between *young* (0.176), *middle-aged* (0.188), and *elderly* (0.192) adults. Lockhart *et al.* (2000b) also reported dynamic frictional demand characteristics between these three age groups on a slippery floor surface (oily vinyl tile) during slip-grip responses (i.e. slip recovery) by measuring adjusted friction utilisation (AFU). A typical kinetic and kinematic profile of a slip-grip response starting from heel contact point on the oily vinyl tile floor surface is shown in figure 5. Heel contact was defined as the point where vertical foot force exceeded 10 N. Initially, as indicated by the horizontal heel position (figure 5(c)), the heel does not slip forward. The horizontal heel velocity decreases (figure 5(b)) as the heel quickly decelerates (figure 5(a)), and both the

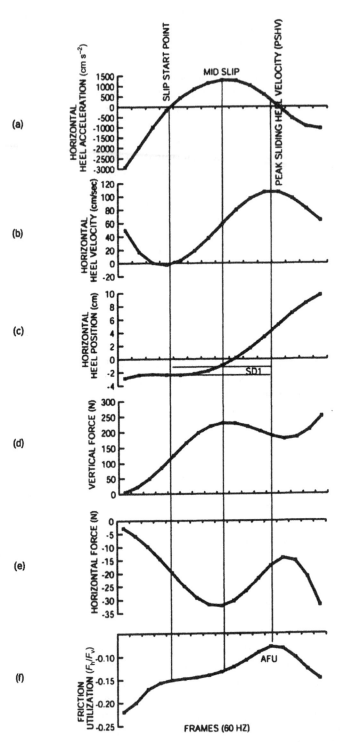

Figure 5. Composite view of the heel dynamics (kinetics and kinematics) during a typical slip-grip response, including adjusted friction utilisation (AFU) on an oily vinyl tile floor surface (Lockhart *et al.* 2000b).

vertical downward force (figure 5(d)) and the horizontal forward force (figure 5(e)) increase. Shortly after heel contact (approx. 60 ms), the heel begins to slip forward (SD1, figure 5(c)). Afterwards, the sliding heel reaches Peak Sliding Heel Velocity (PSHV, figure 5(b)). During this slipping period, the heel accelerates reaching the maximum (figure 5(a)) near the mid-point of the sliding heel velocity profile (figure 5(b)). At this time, both the vertical and the horizontal foot forces decrease. After reaching the maximum heel velocity (approx. 180 ms after heel contact), the sliding heel velocity decreases to a minimum, halting further slipping (figure 5(b)).

AFU is the measured ratio of the horizontal to vertical foot force at the peak sliding heel velocity point (figure 5(f)) and represents the subject's ability to adjust to dynamic frictional requirements during slipping. The significance of this ratio is that it indicates where in the gait profile compensation for a slip is most likely to occur. The AFU of younger individuals (0.074) was adjusted within the dynamic friction requirements (0.08) of the oily vinyl tile floor surface. However, the AFU of middle age (0.10) and older individuals (0.12) was not adjusted within the dynamic friction requirements. Consequently, the result was longer slip distances and increased frequency of falls for these groups.

6. Conclusions

Why do we need to measure slipperiness? Do not current theories on friction mechanisms and biomechanical slip and fall models satisfactorily predict the risks associated with slipperiness? In fact, the underlying mechanisms of slips and falls are not yet fully understood and, therefore, we are not capable of measuring the risks properly. Nevertheless, human-centred approaches for slipperiness measurement do already have many applications. In particular, they are utilized to develop research hypotheses and models to predict workplace risks due to slipping and falling. They are equally important as alternatives to and as means to validate apparatus-based friction measurements, and as practical tools for routine control of slip resistance properties of footwear, anti-skid devices, and floor surfaces.

6.1. *Strengths and weaknesses of human-centred approaches*
Strengths of human-centred approaches over mechanical measurement methodologies are given below:

- the methods are inherently valid for the situation being examined, because human subjects are involved in the experiment, while individual behaviour affects the outcome measures;
- the human factor aspect is included in the analysis and can be partly controlled (e.g. walking speed and cadence, anticipation versus unexpectedly slippery surfaces);
- they allow combining biomechanical measurement data with observations of performance and/or subjective ratings (Strandberg 1985, Strandberg *et al.* 1985, Jung and Schenk 1989, 1990, Hanson *et al.* 1999).

One major concern is that the results obtained with different human-centred methodologies may vary to a great extent, and although it may be a true effect it is sometimes caused by experimental bias; Jung and Fischer (1993) reported in an inter-laboratory study that the outcome of two similar ramp test methods was significantly different for the same test conditions. A possible underlying reason may have been the design of the experimental protocol. Obviously one must strictly control all relevant measurement parameters—such as walking speed and cadence, anticipation of slipperiness, use of safety harness, test environment, sample properties and pre-treatment—in human-centred trials as one needs to do during mechanical friction tests.

Weaknesses of human-centred approaches compared to mechanical slipperiness measurements include the following:

- human-centred experiments are time-consuming and expensive;
- they are mostly suitable for the laboratory environment (field applications are rare, cf. Manning and Jones 1993, Hirvonen *et al.* 1996);
- inter- and intra-individual variation in gait due to anticipation and adaptation to hazards may limit their use (Grönqvist *et al.* 1993);
- learning can affect the measured outcomes, such as magnitude of friction usage and ramp inclination angle or EMG activity changes over repeated trials (Skiba *et al.* 1986, Tang *et al.* 1998);
- a safety harness used to protect subjects from falling can be a confounding factor in many experimental set-ups and may affect the measured outcomes, such as ground reaction forces, required friction, slip velocity and/or distance (Lockhart *et al.* 2000a, b).

6.2. *Differentiation between slipping and falling*

Various definitions of slipping and falling have been used in published studies (cf. Strandberg and Lanshammar 1981, Grönqvist *et al.* 1993, Hirvonen *et al.* 1994, Hanson *et al.* 1999). More precise definitions are needed to discriminate between different outcomes of a slip, for instance, when does a controlled slip (that can be terminated) turn into an uncontrolled slip and fall? The sliding distance of the foot was applied as discriminator between recoverable slips and likely falls by Grönqvist *et al.* (1993). Hanson *et al.* (1999) defined a slip or fall based on each subject's perception; the subjects were asked after each trial whether they subjectively felt 'no slip', 'slip and recovery' or 'slip and fall'. The subjects were also instructed to define a trial as a 'slip and fall' if they required support from a safety harness used to protect them against falling, or if they slid to the end of the force platform that measured the ground reaction force.

Novel experiments are needed to model effective slip recovery and fall avoidance strategies. Pai and Patton (1997) and Pai and Iqbal (1999) used an inverted pendulum model with a foot segment to simulate centre of mass velocity-position constraints and movement termination for balance recovery. In fact, recent studies suggest that in addition to foot displacement during slipping (Brady *et al.* 2000), the movement of the body's centre of mass over the base of support plays a significant role in slip recovery and fall termination (Pai and Patton 1997, Pai and Iqbal 1999, You *et al.* 2001).

6.3. *Safety criteria and thresholds*

Since frictional demands and thresholds for safe friction have been defined differently in different studies, the implications for discriminating safe conditions from potentially hazardous conditions do vary owing to methodological reasons. However, the requirements do also vary by task, subject and gait characteristics as well as by criterion for safe friction such as maximum peak or time-averaged utilised or required friction coefficient. Rarely have suggested frictional demands been related to duration of friction force during stance, which is surprising, since any contact-time-related variation of the utilised or required friction will be omitted, if a maximum peak value is used as the only safety criterion. Consequently, this may lead to misinterpretation of frictional demands that do vary as a function of contact time, particularly in the presence of contaminants between shoes and floor surfaces (Strandberg 1985, Grönqvist *et al.* 2001b).

Current criteria and thresholds for safe friction are still incomplete. Frictional demands and their relation to availability of friction need to be better understood. Environmental consistency (i.e. sufficiently high and steady friction conditions) might be one key factor for improved slip and fall prevention. Perhaps much less can be done to improve human performance or to change work tasks for reducing the frictional demands in the workplace.

7. Future directions of research

For improving the validity and reliability of current risk assessment methods for slips and falls, one must better understand how foot/floor interactions are involved during the events leading to occupational injury and how slip and fall incidents should be simulated. Future directions of research must deal with modelling of basic tribophysical, biomechanical and postural control processes involved in slipping and falling. Slip/fall mechanisms need to be studied for a number of usual and high-risk tasks from walking to manual materials handling, and for different age groups. Events that trigger foot slides and subsequent events during which balance recovery is feasible should be examined in particular. Hypothetical injury mechanisms of falls on the same level and from a higher to a lower level should be tested. The role of slipperiness in the onset and causation of unexpected loading on the musculoskeletal system (especially the low back) is an area of research that has not received sufficient attention.

Objectives for future slips, trips and falls research are detailed below.

(1) To document how anticipated postural controls develop—when does a person predict a potential slipping or tripping condition based on *a priori* knowledge about themselves and the environment, such as proprioceptive and kinesthetic cues, internal situation models characterising spatial relationships of objects, changes in lighting, shadows, floor arrangement, etc? For example, psychophysically, what magnitude of change in lighting intensity constitutes a slip-and-fall warning?

(2) To develop an understanding of how dual-task situations may impede anticipation and adaptation controls—in particular, we want to explain how a person's situation model may influence judgements on temporal and spatial perturbations.

(3) To document how adaptive postural controls develop—what kind of sensory feedback is used in the adaptation—visual, vestibular and proprioceptive feedback? (How do people process information in fall situations?) What kind of feedback makes a person decide to co-activate certain muscles to, for example, gain more leg joint stiffness and maintain an upright torso (i.e. to fight a fall)? What makes a person decide to lower his body for a better fall position to counteract the potential negative consequences of an impending impact?

(4) To understand how postural adjustments associated with voluntary movements are organised for various dual-task situations—how do harmonious motions develop to support primary task performance and balance of gait simultaneously? (This has been labelled physical mode-locking.)

(5) To test the hypothesis that mode-locking between the primary task and gait control facilitates better anticipation and adaptation—if postural adjustments can be made that satisfy both primary and secondary task performance naturally, are cognitive resources off-loaded for improving sensory perception and situation model development to better avoid or address fall events?

(6) To explore mode-locking ability in a younger and an elderly population of healthy subjects and its influence on locomotion under perturbations.

For further exploration of the aetiology of slip and fall accidents on icy surfaces, in particular, and for providing the basis for their prevention, Abeysekera and Gao (2001) presented a systems model, which focuses on an improved understanding of the roles of contributing factors to slip and fall accidents.

The following systems factors were identified.

(1) Footwear (sole) properties including sole material, hardness, roughness, worn/unworn, tread (geometry) design, centre of gravity, anti-slip devices, wearability (weight, height, flexibility, ease of walking, comfort).

(2) Road surface characteristics, covered with ice, snow, contaminants, anti-slip materials, uneven, ascending/descending slope.

(3) Footwear (sole)/road surface interface—the tribological aspect and friction (static, transitional and dynamic friction coefficients).

(4) Human gait biomechanics: muscle strength, postural control, musculoskeletal function, postural reflex and sway, balance capability, acceleration, deceleration, stride length, step length, heel velocity, vertical and horizontal forces.

(5) Human physiological and psychological aspects, i.e. the so-called intrinsic factors, including declines in visual, vestibular, and proprioceptive systems, ageing, perception of slipperiness, information processing, experience, training, diabetes, drug and alcohol usage, unsafe behaviour (rush, reading while walking).

(6) Environment (extrinsic factors): temperature, humidity, snowfall, warm stream, lighting condition, warning and road signs.

Acknowledgements

The authors are grateful to the 'opponent' group members (Mark Redfern, Rakié Cham, Mikko Hirvonen, Håkan Lanshammar, Mark Marpet and Christopher Powers) for their review of the preliminary manuscript presented at the Measurement of Slipperiness symposium in Hopkinton, MA (27–28 July 2000), as well as to all the other reviewers of the draft manuscripts during this process (Gordon Smith, Dal-Ho Son, Chien-Chi Chang and Vincent Ciriello). David A. Winter is gratefully acknowledged for his kind permission to use the figure 2. The authors also wish to thank Lennart Strandberg for valuable amendments to sections 2.3 and 4.3, and Patti Boelsen for her assistance in preparing and proof-reading the manuscript. This manuscript was completed in part during Dr. Grönqvist's tenure as a researcher at the Liberty Mutual Research Center for Safety and Health.

References

ABEYSEKERA, J. and GAO, C. 2001, The identification of factors in the systematic evaluation of slip prevention on icy surfaces, *International Journal of Industrial Ergonomics*, **28**, 303–313.

ALEXANDER, N. B., SHEPHARD, N., MIAN, JU GU and SCHULTZ, A. 1992, Postural control in young and elderly adults when stance is perturbed: Kinematics, *Journal of Gerontology*, **47**, M79–87.

ALLUM, J. H. J., BLOEM, B. R., CARPENTER, M. G., HULLIGER, M. and HADDERS-ALGRA, M. 1998, Proprioceptive control of posture: a review of new concepts, *Gait and Posture*, **8**, 214–242.

ANDRES, R. O., O'CONNOR D. and ENG, T. 1992, A practical synthesis of biomechanical results to prevent slips and falls in the workplace, in S. Kumar (ed.), *Advances in Industrial Ergonomics and Safety IV* (London: Taylor & Francis), 1001–1006.

BERG, K. 1989, Balance and its measure in the elderly: a review, *Physiotherapy in Canada*, **41**, 240–245.

BERG, K. O., WOOD-DAUPHINE, S. L. and MAKI, B. E. 1992a, Measuring balance in the elderly: validation of an instrument, *Canadian Journal of Public Health*, **83**, suppl., 7–11.

BERG, K. O., MAKI, B. E., WILLIAMS, J. I., HOLLIDAY, P. J. and WOOD-DAUPHINE, S. L. 1992b, Clinical and laboratory measures of postural balance in an elderly population, *Archives of Physical Medicine and Rehabilitation*, **73**, 1073–1080.

BOHANNON, R. W. and LEARY, K. M. 1997, Standing balance and function over the course of acute rehabilitation, *Archives of Physical Medicine and Rehabilitation*, **76**, 994–996.

BRADY, R. A., PAVOL, M. J., OWINGS, T. M. and GRABINER, M. D. 2000, Foot displacement but not velocity predicts the outcome of a slip induced in young subjects while walking, *Journal of Biomechanics*, **33**, 803–808.

BRING, C. 1982, Testing of slipperiness: forces applied to the floor and movements of the foot in walking and in slipping on the heel, *Document D5:* 1982 (Stockholm: Swedish Council for Building Research).

BRUCE, M., JONES, C. and MANNING, D. P. 1986, Slip resistance on icy surfaces of shoes, crampons and chains—a new machine, *Journal of Occupational Accidents*, **7**, 273–283.

Brungraber, R. J. 1976, An overview of floor slip-resistance research with annotated bibliography, *National Bureau of Standards (NBS) Technical Note 895* (Washington, DC: US Department of Commerce).

BUCZEK, F. L., CAVANAGH, P. R., KULAKOWSKI, B. T. and PRADHAN, P. 1990, Slip resistance needs of the mobility disabled during level and grade walking, in B. E. Gray (ed.), *Slips, Stumbles, and Falls: Pedestrian Footwear and Surfaces*, ASTM STP 1103 (Philadelphia, PA: American Society for Testing and Materials), 39–54.

BULAJIC-KOPIAR, M. 2000, Seasonal variation in incidence of fractures among elderly people, *Injury Prevention*, **6**, 16–19.

CARLSÖÖ, S. 1962, A method for studying walking on different surfaces, *Ergonomics*, **5**, 271–274.

CHAM, R., MUSOLINO, M. and REDFERN, M. S. 2000, Heel contact dynamics during slip events, Proceedings of the IEA 2000/HFES 2000 Congress, Vol. 4, *Safety and Health, Aging* (San Diego, CA: International Ergonomics Association), 514–517.

CHANG, W. R., GRÖNQVIST, R., LECLERCQ, S., MYUNG, R., MAKKONEN, L., STRANDBERG, L., BRUNGRABER, R., MATTKE, U. and THORPE, S. 2001a, The role of friction in the measurement of slipperiness, Part I: Friction mechanisms and definitions of test conditions, *Ergonomics*, **44**, 1217–1232.

CHANG, W. R., LECLERCQ, S., GRÖNQVIST, R., BRUNGRABER, R., MATTKE, U., STRANDBERG, L., THORPE, S., MYUNG, R., MAKKONEN, L. and COURTNEY, T. K. 2001b, The role of friction in the measurement of slipperiness, Part II: Survey of friction measurement devices, *Ergonomics*, **44**, 1233–1261.

CHIOU, S., BHATTACHARYA, A. and SUCCOP, P. A. 1996, Effect of workers' shoe wear on objective and subjective assessment of slipperiness, *American Industrial Hygiene Association Journal*, **57**, 825–831.

CHIOU, S., BHATTACHARYA, A. and SUCCOP, P. A. 2000, Evaluation of workers' perceived sense of slip and effect of prior knowledge of slipperiness during task performance on slippery surfaces, *American Industrial Hygiene Association Journal*, **61**, 492–500.

CHIU, J. and ROBINOVITCH, N. 1998, Prediction of upper extremity impact forces during falls on the outstretched hand, *Journal of Biomechanics*, **31**, 1169–1176.

CHRISTINA, K. A., OKITA, N., OWENS, D. A. and CAVANAGH, P. R. 2000, Stair descent in the elderly: effects of visual conditions on foot-stair interaction, *Proceedings of the IEA 2000/HFES 2000 Congress, Vol. 4, Safety and Health, Aging* (San Diego, CA: International Ergonomics Association), 510–513.

CIRIELLO, V. M., McGORRY, R. W. and MARTIN, S. E. 2001, Maximum acceptable horizontal and vertical forces of dynamic pushing on high and low coefficient of friction floors, *International Journal of Industrial Ergonomics*, **27**, 1–8.

COHEN, H. H. and COHEN, M. D. 1994a, Psychophysical assessment of the perceived slipperiness of floor tile surfaces in a laboratory setting, *Journal of Safety Research*, **25**, 19–26.

COHEN, H. H. and COHEN, M. D. 1994b, Perceptions of walking surface slipperiness under realistic conditions, utilizing a slipperiness rating scale, *Journal of Safety Research*, **25**, 27–31.

DANION, F., BONNARD, M. and PAILHOUS, J. 1997, Intentional on-line control of propulsive forces in human gait, *Experimental Brain Research*, **116**, 525–538.

DELANGE, A. and GRÖNQVIST, R. 1997, Bridging the gap between mechanical and biomechanical slip test methods, in P. Seppälä, T. Luopajärvi, C.-H. Nygård and M. Mattila (eds), *From Experience to Innovation*, IEA'97, Vol. 3 (Helsinki: Finnish Institute of Occupational Health), 383–385.

EKDAHL, C., JARNLO, G. B. and ANDERSSON, S. I. 1989, Standing balance in healthy subjects: Evaluation of a quantitative test battery on a force platform, *Scandinavian Journal of Rehabilitation Medicine*, **21**(4), 187–195.

ENDSLEY, M. R. 1995, Towards a new paradigm for automation: designing for situation awareness, *Proceedings of the 6th IFAC/IFIP/IFORS/IEA Symposium on Analysis, Design and Evaluation of Man-Machine Systems*, Vol. 1, 365–370.

ENG, J. J, WINTER, D. A. and PATLA, A. E. 1994, Strategies for recovery from a trip in early and late swing during human walking, *Experimental Brain Research*, **102**, 339–349.

ENGSTRÖM, J. and BURNS, P. C. 2000, Psychophysical methods for quantifying opinions and preferences, in P. T. McCabe, M. A. Hanson and S. A. Robertson (eds), *Contemporary Ergonomics 2000* (London: Taylor & Francis), 234–238.

FATHALLAH, F. A. and COTNAM, J. P. 2000, Maximum forces sustained during various methods of exiting commercial tractors, trailers and trucks, *Applied Ergonomics*, **31**, 25–33.

FATHALLAH, F. A., GRÖNQVIST, R. and COTNAM, J. P. 2000, Estimated slip potential on icy surfaces during various methods of exiting commercial tractors, trailers and trucks, *Safety Science*, **36**, 69–81.

FOTHERGILL, J., O'DRISCOLL, D. and HASHEMI, K. 1995, The role of environmental factors in causing injury through falls in public places, *Ergonomics*, **38**, 220–223.

FUGL-MEYER, A. R., JÄÄSKÖ, L., LEYMAN, O., OLSSON, S. and STEGLIND, S. 1975, The post-stroke hemiplegic patient. A method for evaluation of physical performance, *Scandinavian Journal of Rehabilitation Medicine*, **7**, 13–31.

GARD, G. and LUNDBORG, G. 2000, Pedestrians on slippery surfaces during winter—methods to describe the problems and tests of anti-skid devices, *Accident Analysis & Prevention*, **32**, 455–460.

GARD, G. and LUNDBORG, G. 2001, Test of anti-skid devices on five different slippery surfaces, *Accident Analysis & Prevention*, **33**, 1–8.

GRIEVE, D. W. 1979, Environmental constraints on the static exertion of force: PSD analysis in task-design, *Ergonomics*, **22**, 1165–1175.

GRIEVE, D. W. 1983, Slipping due to manual exertion, *Ergonomics*, **26**, 61–72.

GRÖNQVIST, R. 1999, Slips and falls, in S. Kumar (ed.), *Biomechanics in Ergonomics* (London: Taylor & Francis), 351–375.

GRÖNQVIST, R. and ROINE, J. 1993, Serious occupational accidents caused by slipping, in R. Nielsen and K. Jorgensen (eds), *Advances in Industrial Ergonomics and Safety V* (London: Taylor & Francis), 515–519.

GRÖNQVIST, R., HIRVONEN, M. and MATZ, S. 2001b, Walking safety and contact time related variation in shoe-floor traction, *Proceedings of the International Conference on Computer-Aided Ergonomics and Safety*, 29 July–1 August, Maui, Hawaii.

GRÖNQVIST, R., HIRVONEN, M. and TOHV, A. 1999, Evaluation of three portable floor slipperiness testers, *International Journal of Industrial Ergonomics*, **25**, 85–95.

GRÖNQVIST, R., HIRVONEN, M. and TUUSA, A. 1993, Slipperiness of the shoe-floor interface: comparison of objective and subjective assessments, *Applied Ergonomics*, **24**, 258–262.

GRÖNQVIST, R., ROINE, J., JÄRVINEN, E. and KORHONEN, E. 1989, An apparatus and a method for determining the slip resistance of shoes and floors by simulation of human foot motions, *Ergonomics*, **32**, 979-995.

GRÖNQVIST, R., CHANG, W. R., COURTNEY, T. K., LEAMON, T. B., REDFERN, M. S. and STRANDBERG, L. 2001a, Measurement of slipperiness: fundamental concepts and definitions, *Ergonomics*, **44**, 1102–1117.

HANSON, J. P., REDFERN, M. S. and MAZUMDAR, M. 1999, Predicting slips and falls considering required and available friction, *Ergonomics*, **42**, 1619–1633.

HARPER, F. C., WARLOW, W. J. and CLARKE, B. L. 1967, The forces applied to the floor by the foot in walking. Part II. Walking on a slope. Part III. Walking on stairs, *National Building Studies*, Research Paper 32 (London: Ministry of Technology).

HIRVONEN, M., LESKINEN, T., GRÖNQVIST, R. and SAARIO, J. 1994, Detection of near accidents by measurement of horizontal acceleration of the trunk, *International Journal of Industrial Ergonomics*, **14**, 307–314.

HIRVONEN, M., LESKINEN, T., GRÖNQVIST, R., VIIKARI-JUNTURA, E. and RIIHIMÄKI, H. 1996, Occurrence of sudden movements at work, *Safety Science*, **24**, 77–82.

HONKANEN, R. 1982, The role of slippery weather in accidental falls, *Journal of Occupational Accidents*, **4**, 257–262.

HONKANEN, R. 1983, The role of alcohol in accidental falls, *Journal of Studies on Alcohol*, **44**, 231–245.

HORAK, F. B. 1997, Clinical assessment of balance disorders, *Gait and Posture*, **6**, 76–84.

HSIAO, E. T. and ROBINOVITCH, S. N. 1998, Common protective movements govern unexpected falls from standing height, *Journal of Biomechanics*, **31**, 1–9.

HSIAO, E. T. and ROBINOVITCH, S. N. 1999, Biomechanical influences on balance recovery by stepping, *Journal of Biomechanics*, **32**, 1099–1106.

ISAACS, B. 1985, Clinical and laboratory studies of falls in old people, *Clinical Geriatric Medicine*, **1**, 513.

JAMES, D. I. 1980, A broader look at pedestrian friction, *Rubber Chemistry and Technology*, **53**, 512–541.

JAMES, D. I. 1983, Rubber and plastics in shoes and flooring: the importance of kinetic friction, *Ergonomics*, **26**, 83–99.

JOHANSSON, R. and MAGNUSSON, M. 1991, Human postural dynamics, *Biomedical Engineering*, **18**, 413–437.

JONES, M. R. 1976, Time, our lost dimension: toward a new theory of perception, attention, and memory, *Psychological Review*, **83**, 323–355.

JUNG, K. and FISCHER, A. 1993, Methods for checking the validity of technical test procedures for the assessment of slip resistance of footwear, *Safety Science*, **16**, 189–206.

JUNG, K. and RÜTTEN, A. 1992, Entwicklung eines Verfahrens zur Prüfung der Rutschemmung von Bodenbelägen für Arbeitsräume, Arbeitsbereich und Verkehrswege, *Zentralblatt für Arbeitsmedizin, Arbeitsschutz, Prophylaxe und Ergonomie*, **42**(6), 227–235 (in German with English summary).

JUNG, K. and SCHENK, H. 1989, Objektivierbarkeit und Genauigkeit des Begehungsverfahrens zur Ermittlung der Rutschemmung von Bodenbelägen, Zentralblatt für Arbeitsmedizin, Arbeitsschutz, Prophylaxe und Ergonomie, 39(8), 221–228 (in German with English summary).

JUNG, K. and SCHENK, H. 1990, Objektivierbarkeit und Genauigkeit des Begehungsverfahrens zur Ermittlung der Rutschemmung von Schuhen, *Zentralblatt für Arbeitsmedizin, Arbeitsschutz, Prophylaxe und Ergonomie*, **40**(3), 70–78 (in German with English summary).

KINOSHITA, H. 1985, Effects of loads and carrying systems on selected biomechanical parameters describing walking gait, *Ergonomics*, **28**, 1347–1362.

KROEMER, K. H. E. 1974, Horizontal push and pull forces—exertable when standing in working positions on various surfaces, *Applied Ergonomics*, **5**, 94–102.

LANSHAMMAR, H. and STRANDBERG, L. 1983, Horizontal floor reaction forces and heel movements during the initial stance phase, in H. Matsui and K. Kobayashi (eds), *Biomechanics VIII* (Baltimore, MD: University Park Press), 1123–1128.

LANSHAMMAR, H. and STRANDBERG, L. 1985, Assessment of friction by speed measurement during walking in a closed path, in D. A. Winter, R. W. Norman, R. P. Wells, K. C. Hayes and A. E. Patla (eds), *Biomechanics IX-B* (Champaign, IL: Human Kinetics Publishers), 72–75.

LEAMON, T. B. and LI, K.-W. 1990, Microslip length and the perception of slipping, paper presented at the 23rd International Congress on Occupational Health, 22–28 September, Montreal, Canada.

LEAMON, T. B. and MURPHY, P. L. 1995, Occupational slips and falls: more than a trivial problem, *Ergonomics*, **38**, 487–498.

LEAMON, T. B. and SON, D. H. 1989, The natural history of a microslip, in A. Mital (ed.), *Advances in Industrial Ergonomics and Safety I* (London: Taylor & Francis), 633–638.

LECLERCQ, S. 1999, The prevention of slipping accidents: a review and discussion of work related to the methodology of measuring slip resistance, *Safety Science*, **31**, 95–125.

LINDBERG, P. Å. and STÅLHANDSKE, P. 1981, Simulation models of slipping and falling human bodies (in Swedish: Simuleringsmodeller för halkning), Master's thesis, Royal Institute of Technology, Stockholm.

LLEWELLYN, M. G. A. and NEVOLA, V. R. 1992, Strategies for walking on low-friction surfaces, in W. A. Lotens and G. Havenith (eds), *Proceedings of the Fifth International Conference on Environmental Ergonomics*, Maastricht, The Netherlands, 156–157.

LOCKHART, T. E. 1997, The ability of elderly people to traverse slippery walking surfaces, *Proceedings of the Human Factors and Ergonomics Society 41st Annual Meeting*, Vol. 1 (Albuquerque: Human Factors and Ergonomics Society), 125–129.

LOCKHART, T. E., SMITH, J. L., WOLDSTAD, J. C. and LEE, P. S. 2000a, Effects of musculoskeletal and sensory degradation due to aging on the biomechanics of slips and falls, *Proceedings of the IEA/HFES Congress*, Vol. 5, *Industrial Ergonomics* (San Diego, CA: International Ergonomics Association), 83–86.

LOCKHART, T. E., WOLDSTAD, J. C., SMITH, J. L. and HSIANG, S. M. 2000b, Prediction of falls using a robust definition of slip distance and adjusted required coefficient of friction, *Proceedings of the IEA/HFES Congress*, Vol. 4, *Safety and Health, Aging* (San Diego, CA: International Ergonomics Association), 506–509.

LOCKHART, T. E., WOLDSTAD, J. C., SMITH, J. L. and RAMSEY, J. D. 2002, Effects of age-related sensory degradation on perception of floor slipperiness and associated slip parameters, *Safety Science*, **40**(7–8), 689–703.

LUND, J. 1984, Accidental falls at work, in the home and during leisure activities, *Journal of Occupational Accidents*, **6**, 181–193.

MAKI, B. E. and MCILROY, W. E. 1997, The role of limb movements in maintaining upright stance: the 'change-in-support' strategy, *Physical Therapy*, **77**, 488–507.

MALMIVAARA, A., HELIÖVAARA, M., KNEKT, P., REUNANEN, A. and AROMAA, A. 1993, Risk factors for injurious falls leading to hospitalization or death in a cohort study of 19,500 adults, *American Journal of Epidemiology*, **138**, 384–394.

MANNING, D. P. and JONES, C. 1993, A step towards safe walking, *Safety Science*, **16**, 207–220.

MANNING, D. P., JONES, C. and BRUCE, M. 1991, A method of ranking the grip of industrial footwear on water wet, oily and icy surfaces, *Safety Science*, **M14**, 1–12.

MANNING, D. P., AYERS, I., JONES, C., BRUCE, M. and COHEN, K. 1988, The incidence of underfoot accidents during 1985 in a working population of 10,000 Merseyside people, *Journal of Occupational Accidents*, **10**, 121–130.

MATHIAS, S., NAYAK, U. S. L. and ISAACS, B. 1986, Balance in the elderly patient. The get up and go test, *Archives of Physical and Medicical Rehabilitation*, **67**, 387.

MCILROY, W. E. and MAKI, B. E. 1999, The control of lateral stability during rapid stepping reactions evoked by antero-posterior perturbation: does anticipatory control play a role? *Gait and Posture*, **9**, 190–198.

MCVAY, E. J. and REDFERN, M. S. 1994, Rampway safety: foot forces as a function of rampway angle, *American Industrial Hygiene Association Journal*, **55**, 626–634.

MERRILD, U. and BAK, S. 1983, An excess of pedestrian injuries in icy conditions: A high-risk fracture group—elderly women, *Accident Analysis & Prevention*, **15**, 41–48.

MOORE, D. F. 1972, The friction and lubrication of elastomers, in G. V. Raynor (ed.), *International Series of Monographs on Materials Science and Technology*, Vol. 9 (Oxford: Pergamon Press).

MORACH, B. 1993, Quantifierung des Ausgleitvorganges beim menschlichen Gang unter besonderer Berücksichtigung der Aufsetzphases des Fusses, Fachbereich Sicherheitstechnik der Bergischen Universität—Gesamthochschule Wuppertal, Doctoral dissertation (in German).

MORASSO, P. G., SPADA, G. and CAPRA, R. 1999, Computing the COM from the COP in postural sway movements, *Human Movement Science*, **18**, 759 – 767.

MYUNG, R. and SMITH, J. L. 1997, The effect of load carrying and floor contaminants on slip and fall parameters, *Ergonomics*, **40**, 235 – 246.

MYUNG, R., SMITH, J. L. and LEAMON, T. B. 1992, Slip distance for slip/fall studies, in S. Kumar (ed.), *Advances in Industrial Ergonomics and Safety IV* (London: Taylor & Francis), 983 – 987.

MYUNG, R., SMITH, J. L. and LEAMON, T. B 1993, Subjective assessment of floor slipperiness, *International Journal of Industrial Ergonomics*, **11**, 313 – 319.

NAGATA, H. 1989, The methodology of insuring the validity of a slip-resistance meter, *Proceedings of the International Conference on Safety* (Tokyo: Metropolitan Institute of Technology), 33 – 38.

NAGATA, H. 1993, Fatal and non-fatal falls—a review of earlier articles and their developments, *Safety Science*, **16**, 379 – 390.

NASHNER, L. M. 1983, Analysis of movement control in man using the movement platform, in E. Desmedt (ed.), *Motor Control Mechanisms in Health and Disease* (New York: Raven Press), 607 – 619.

NASHNER, L. M. 1985, Conceptual and biomechanical models of postural control, in M. Igarashi and F. Black (eds), *Vestibular and Visual Control of Posture and Locomotor Equilibrium* (Basel: Karger), 1 – 8.

PAI, Y.-C. and PATTON, J. 1997, Center of mass velocity-position predictions for balance control, *Journal of Biomechanics*, **30**, 347 – 354.

PAI, Y.-C. and IQBAL, K. 1999, Simulated movement termination for balance recovery: can movement strategies be sought to maintain stability in the presence of slipping or forced sliding? *Journal of Biomechanics*, **32**, 779 – 786.

PATLA, A. E. 1991, Visual control of human locomotion, in A. E. Patla (ed.), *Adaptability of Human Gait: Implications for the Control of Locomotion* (Amsterdam: Elsevier/North-Holland), 55 – 97.

PERKINS, P. J. 1978, Measurement of slip between the shoe and ground during walking, in C. Anderson and J. Senne (eds), *Walkway Surfaces: Measurement of Slip Resistance*, ASTM STP 649 (Baltimore, MD: American Society for Testing and Materials), 71 – 87.

PERKINS, P. J. and WILSON, P. 1983, Slip resistance testing of shoes—new developments, *Ergonomics*, **26**, 73 – 82.

PODSIADLO, D. and RICHARDSSON, S. 1991, The timed up and go: a test of basic functional mobility for frail elderly persons, *Journal of American Geriatric Society*, **39**, 142 – 148.

PYYKKÖ, I., AALTO, H. and STARCK, J. 1988, Postural control in bilateral vestibular disease, in C.-F. Claussen, M. V. Kirtane and K. Schlitter (eds), *Vertigo, Nausea, Tinnitus and Hypoacusia in Metabolic Disorders* (Amsterdam: Elsevier), 473 – 476.

PYYKKÖ, I., JÄNTTI, P. and AALTO, H. 1990, Postural control in elderly subjects, *Age and Aging*, **19**, 215 – 221.

REDFERN, M. S. and DIPASQALE, J. 1997, Biomechanics of descending ramps, *Gait and Posture*, **6**, 119 – 125.

REDFERN, M. S. and RHOADES, T. P. 1996, Fall prevention in industry using slip resistance testing, in A. Bhattacharya and J. D. McGlothlin (eds), Occupational Ergonomics, Theory and Applications (New York/Basle/Hong Kong: Marcel Dekker), 463 – 476.

REDFERN, M. S. and SCHUMAN, T. 1994, A model of foot placement during gait, *Journal of Biomechanics*, **27**, 1339 – 1346.

REDFERN, M., CHAM, R., GIELO-PERCZAK, K., GRÖNQVIST, R., HIRVONEN, M., LANSHAMMAR, H., MARPET, M., PAI, Y.-C. and POWERS, C. 2001, Biomechanics of slips, *Ergonomics*, **44**, 1138 – 1166.

ROBINOVITCH, S. N., HSIAO, E., KEARNY, M. and FRENK, V. 1996, Analysis of movement strategies during unexpected falls, presentation at the 20th Annual Meeting of the American Society of Biomechanics, October 17–19, Atlanta, Georgia.

SCHEIL, M. and WINDHÖVEL, U. 1994, Instationäre Reibzahlmessung mit dem Messverfahren nach Manning, *Zeitschrift für Arbeitswissenschaft*, **20**, 177–181 (in German with English summary).

SKIBA, R. 1988, Sicherheitsgrenzwerte zur Vermeidung des Ausgleitens auf FussbÖden. *Zeitschrift für Arbeitswissenschaft*, **14**, 47–51 (in German with English summary).

SKIBA, R., BONEFELD, X. and MELLWIG, D. 1983, Voraussetzung zur Bestimmung der Gleitsicherheit beim menschlichen Gang, *Zeitschrift für Arbeitswissenschaft*, **9**, 227–232 (in German with English summary).

SKIBA, R., WIEDER, R. and CZIUK, N. 1986, Zum Erkinntniswert von Reibzahlmessung durch Begehen einer neigbaren Ebene, *Kautschuk + Gummi Kunststoffe*, **39**, 907–911 (in German with English summary).

SKIBA, R., WORTMANN, H. R. and MELLWIG, D. 1985, Bewegungsabläufe und Kräfte beim Treppenaufstieg un -abstieg aus der Sicht der Gleitsicherheit, *Zeitschrift für Arbeitswissenschaft*, **11**, 97–100 (in German with English summary).

SOROCK, G. S. 1988, Falls among the elderly: epidemiology and prevention, *American Journal of Preventive Medicine*, **4**, 282–288.

STRANDBERG, L. 1983, On accident analysis and slip-resistance measurement, *Ergonomics*, **26**, 11–32.

STRANDBERG, L. 1985, The effect of conditions underfoot on falling and overexertion accidents, *Ergonomics*, **28**, 131–147.

STRANDBERG, L. and LANSHAMMAR, H. 1981, The dynamics of slipping accidents, *Journal of Occupational Accidents*, **3**, 153–162.

STRANDBERG, L., HILDESKOG, L. and OTTOSON, A.-L. 1985, Footwear friction assessed by walking experiments, *VTIrapport 300 A*, Väg- och trafikinstitutet, Linköping.

SWENSEN, E., PURSWELL, J., SCHLEGEL, R. and STANEVICH, R. 1992, Coefficient of friction and subjective assessment of slippery work surfaces, *Human Factors*, **34**, 67–77.

TANG, P.-F., WOOLLACOTT, M. H. and CHONG, R. K. Y. 1998, Control of reactive balance adjustments in perturbed walking: roles of proximal and distal postural muscle activity, *Experimental Brain Research*, **119**, 141–152.

TEMPLER, J., ARCHEA, J. and COHEN, H. H. 1985, Study of factors associated with risk of work-related stairway falls, *Journal of Safety Research*, **16**, 183–196.

TINETTI, M. E., WILLIAMS, T. F. and MAYEWSKI, R. 1986, Fall risk index for elderly patients based on number of chronic disabilities, *American Journal of Medicine*, **80**, 429–434.

TISSERAND, M. 1985, Progress in the prevention of falls caused by slipping, *Ergonomics*, **28**, 1027–1042.

VAUGHAN, C. L. 1984, Biomechanics of running gait, *CRC Critical Reviews in Biomedical Engineering*, **12**, Issue 1 (Boca Raton: CRC Press), 1–48.

WALLER, J. A. 1978, Falls among the elderly—human and environmental factors, *Accident Analysis & Prevention*, **10**, 21–33.

WINTER, D. A. 1991, *The Biomechanics and Motor Control of Human Gait: Normal, Elderly, and Pathological*, 2nd edn (Ontario: University of Waterloo), 143.

WINTER, D. A. 1995, *ABC Anatomy, Biomechanics and Control of Balance during Standing and Walking* (Ontario: University of Waterloo).

WINTER, D. A., PATLA, A. F., PRINCE, F., ISHAC, M. and GIELO-PERCZAK, K. 1998, Stiffness control of balance in quiet standing, *Journal of Neurophysiology*, **80**, 1211–1221.

YOSHIOKA, M., ONO, H., KAWAMURA, S. and MIYAKI, M. 1978, On slipperiness of building floors—fundamental investigation for scaling of slipperiness, *Report of the Research Laboratory of Engineering Materials, No. 3* (Tokyo: Tokyo Institute of Technology), 129–134.

YOSHIOKA, M., ONO, H., SHINOHARA, M., KAWAMURA, S., MIYAKI, M. and KAWATA, A. 1979, Slipperiness of building floors, *Report of the Research Laboratory of Engineering Materials, No. 4* (Tokyo: Tokyo Institute of Technology), 140–157.

YOU, J.-Y., CHOU, Y.-L., LIN, C.-J. and SU, F.-C. 2001, Effect of slip on movement of body center of mass relative to base of support, *Clinical Biomechanics*, **16**, 167–173.

ZHAO, Y., UPADHYAYA, S. K. and KAMINAKA, M. S. 1987, Foot-ground forces on sloping ground when lifting, *Ergonomics*, **30**, 1671–1687.

CHAPTER 5

The role of surface roughness in the measurement of slipperiness

Wen-Ruey Chang†*, In-Ju Kim‡, Derek P. Manning§ and Yuthachai Bunterngchit¶

†Liberty Mutual Research Center for Safety and Health, 71 Frankland Road, Hopkinton, MA 01748, USA

‡School of Exercise and Sport Science, The University of Sydney, Lidcombe, NSW 2141, Australia

§341 Liverpool Road, Birkdale, Southport, Merseyside PR8 3DE, UK

¶Department of Industrial Engineering, King Mongkut's Institute of Technology North Bangkok, Bangsue, Bangkok 10800, Thailand

Keywords: Surface roughness; Slipperiness; Slip resistance.

Surface roughness has been shown to have substantial effects on the slip resistance between shoe heels and floor surfaces under various types of walking environments. This paper summarises comprehensive views of the current understanding on the roles of surface roughness on the shoe and floor surfaces in the measurement of slipperiness and discusses promising directions for future research. Various techniques and instruments for surface roughness measurements and related roughness parameters are reviewed in depth. It is suggested that a stylus-type profilometer and a laser scanning confocal microscope are the preferred instruments for surface roughness measurements in the field and laboratory, respectively. The need for developing enhanced methods for reliably characterising the slip resistance properties is highlighted. This could be based on the principal understanding of the nature of shoe and floor interface and surface analysis techniques for characterising both surfaces of shoe and floor. Therefore, surface roughness on both shoe and floor surfaces should be measured and combined to arrive at the final assessment of slipperiness. While controversies around the friction measurement for slipperiness assessment still remain, surface roughness measurement may provide an objective alternative to overcoming the limitations of friction measurements.

1. Introduction

There is a general awareness that smooth floor surfaces are slippery, especially when wet, and that rougher surfaces are safer. It was only in the last two decades that scientific research revealed that the microscopic roughness of floor surfaces and shoe solings has a profound effect on underfoot friction. The microscopic roughness must be distinguished from the moulded surface pattern. A macroscopic profile assists penetration of a film of liquid on the floor enabling the minute peaks, known as asperities, of footwear and floors to make contact.

---

*Author for correspondence. e-mail: Wen.Chang@LibertyMutual.com

Although over 70 machines have been invented to measure slip resistance (Strandberg 1985), none of them accurately represents the motion of a human foot and, at present, there is no generally accepted method of measuring slipperiness. The complexities of the forces involved in human locomotion and the infinite variety of the shoe/floor/contaminant interfaces may prove to be insuperable obstacles to the successful design of a method of measuring slipperiness. The research on surface roughness offers another pathway to the design of slip-resistant footwear and floors. It may be possible in the future to rate slip resistance in terms of the parameters and magnitude of surface roughness as an alternative to the measurement of the coefficient of friction (COF). At the very least, surface roughness must be measured in association with the COF because it has been proved that the COF is highly dependant on the roughness values and parameters. In addition, the results of surface roughness measurements could be very valuable to floor and shoe manufacturers and end users to properly improve or select floor and shoe products.

Leonardo da Vinci may have been the first scientist to discover that when one object slides on another, friction is influenced by surface roughness (MacCurdy 1938). The French engineer Amontons, who discovered two of the laws of friction in 1699, also realised that the touching surfaces were not flat and asserted that the roughness on surfaces must interlock with each other (Kennaway 1970). Coulomb suggested in 1781 that friction was due to the interlocking of the surface asperities and represented the work of lifting the load over the summits of the asperities (Bowden and Tabor 1964). Hunter (1930) indicated that roughness of both the shoe and floor surfaces had a great influence on COF. Sigler *et al.* (1948) noted that good anti-slip properties under wet conditions are usually associated with the rough particles that project through the film of water and thus prevent its action as a lubricant. Bowden and Tabor (1964) mentioned that hills and valleys are present which are large compared with the size of a molecule even on carefully polished surfaces. If two solids are placed in contact, the upper surface will be supported on the summits of the irregularities, and large areas of the surfaces will be separated by a distance greater than the molecular range of action.

Jung and Reidiger (1982) appear to be the first workers to measure surface roughness of floors. Using a photo-optical method, they found that the surface roughness related well with both objective and subjective assessments of slip resistance and that the parameter R_t, the peak to valley height, correlated well with COF. They measured static friction by sliding a weighted leather or rubber sample down an inclined plane covered with the chosen floor material lubricated by engine oil. Manning *et al.* (1983) reported that the slip resistance of footwear solings on oily surfaces was dependant on surface roughness assessed by naked eye observation. Harris and Shaw (1988) reported that there was strong correlation between users' opinions of floor safety and roughness levels of floor surfaces in R_{tm}, the average peak to valley distance, measured with the Rank Taylor Hobson (currently known as Taylor Hobson Precision, Leicester, UK) Surtronic 10, a pocket sized, portable instrument.

The objective of the paper is to summarise the current understanding of surface roughness measurements related to the measurements of slipperiness and discuss promising directions for future research. Understanding surface roughness in slip resistance can also help us to understand friction and wear mechanisms involved at the shoe and floor interface.

2. Definitions of commonly used surface roughness parameters

Various surface roughness parameters can be generated from a surface profile to represent its geometric characteristics. Definitions of numerous surface roughness parameters are found in several international standards and also the manuals of profilometers widely used throughout the world (British Standards Institution 1988, Whitehouse 1994, Rank Taylor Hobson Ltd 1996, International Organization of Standardization 1998).

High-pass filtering is performed on the measured profile with a proper selection of a filtering length (the cut-off length, l_r) to obtain the surface roughness profile, also known as the surface heights. The surface roughness parameters are calculated from the filtered roughness profile. A typical surface roughness profile is shown in figure 1. The evaluation length l_n (the assessed length) usually consists of an integral multiplication of the cut-off length. The height of the assessed profile at any position x can be obtained from a general function $Z(x)$ to mathematically describe the surface. However, surface profiles measured with a profilometer are typically digitised. Discrete points (x_i, $i = 1$, ..., n) with an equal increment Δx and the corresponding surface heights (z_i, $i = 1$, ..., n) are used instead to describe the surface profiles. These points are related by

$$x_i = x_1 + (i - 1) \times \Delta x \tag{1}$$

$$z_i = Z(x_i) \tag{2}$$

Typically the mean line of the surface heights is drawn such that $\sum_{i=1}^{n} z_i = 0$ as shown in figure 1 with $z = 0$ as the mean line.

Commonly used surface roughness parameters are defined below.

R_a is the arithmetical average of surface heights, also known as the centre line average of surface heights (CLA), and can be calculated as

$$R_a = \frac{1}{n} \sum_{i=1}^{n} |z_i| \text{ or } R_a = \frac{1}{l_n} \int_0^{l_n} |Z(x)| dx \tag{3}$$

R_q is defined as the root mean square of surface heights, i.e.

$$R_q = \sqrt{\frac{1}{n} \sum_{i=1}^{n} z_i^2} \text{ or } R_q = \sqrt{\frac{1}{l_n} \int_0^{l_n} Z^2(x) dx} \tag{4}$$

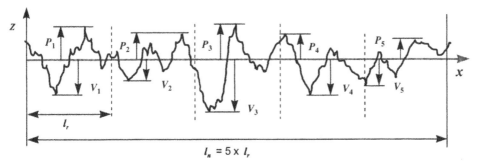

Figure 1. A typical surface roughness profile.

Two parameters related to the shape of surfaces are the skewness, R_{sk}, and the kurtosis, R_{ku}, in which the shape and hence the distribution of heights are described by numerical values. The skewness of surface heights R_{sk}, as shown in figure 2, is defined as

$$R_{sk} = \frac{1}{nR_q^3}\sum_{i=1}^{n} z_i^3 \text{ or } R_{sk} = \frac{1}{l_n R_q^3}\int_0^{l_n} Z^3(x)\mathrm{d}x \qquad (5)$$

The kurtosis of surface heights R_{ku}, as shown in figure 3, is defined as

$$R_{ku} = \frac{1}{nR_q^4}\sum_{i=1}^{n} z_i^4 \text{ or } R_{ku} = \frac{1}{l_n R_q^4}\int_0^{l_n} Z^4(x)\mathrm{d}x \qquad (6)$$

These parameters may provide a quantitative measure of a clear visual appearance of surface profiles. Profiles in which peaks are dominant have positive skewness and

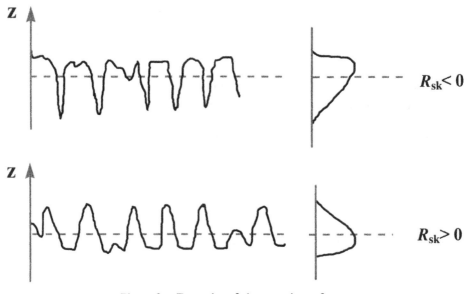

Figure 2.   Examples of skewness in surfaces.

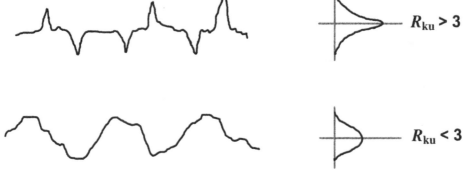

Figure 3.   Examples of kurtosis in surfaces.

profiles dominated by valleys have negative skewness. Kurtosis is a measure of the peakedness or spikiness of a surface profile. For a surface with a normal distribution of surface heights, R_{sk} is zero and R_{ku} is 3. If $R_{ku} < 3$, the surface profile consists of mostly rounded peaks and rounded valleys as shown in figure 3. As R_{ku} increases above 3, the shapes of its peaks and valleys become less rounded.

In an example shown in figure 1, there are five cut-off lengths in the total evaluation length, and the highest peak and the lowest valley in each cut-off length are identified as P_i and V_i where $i = 1$ to 5 and all P_i and V_i have positive values. R_p is the maximum height of the profile above the mean line within the assessed length and can be obtained from the maximum of P_i. R_v is the maximum depth of the profile below the mean line within the assessed length and can be obtained from the maximum V_i. R_t is the maximum peak to valley height in the assessed length and is equal to the sum of R_p and R_v. R_{pm} is the average of the maximum height above the mean line in each cut-off length and R_{vm} is the average of the maximum depth below the mean line in each cut-off length. R_{pm} and R_{vm} can be obtained from

$$R_{pm} = \frac{\sum_{i=1}^{5} P_i}{5} \tag{7}$$

$$R_{vm} = \frac{\sum_{i=1}^{5} V_i}{5} \tag{8}$$

R_{tm} is the average of peak to valley height in each cut-off length and is equal to the sum of R_{pm} and R_{vm}, i.e.

$$R_{tm} = R_{pm} + R_{vm} \tag{9}$$

R_{tm} is also known as R_z in DIN standards.

The local surface slope can be calculated in the discrete form as

$$\frac{dz_i}{dx} = \frac{1}{60\Delta x}(z_{i+3} - 9z_{i+2} + 45z_{i+1} - 45z_{i-1} + 9z_{i-2} - z_{i-3}) \tag{10}$$

Δ_a and Δ_q are the arithmetical mean and root mean square of surface slope, respectively, and can be obtained in the form of

$$\Delta_a = \frac{1}{n}\sum_{i=1}^{n} |\frac{dz_i}{dx}| \text{ or } \Delta_a = \frac{1}{l_n}\int_0^{l_n} |\frac{dZ}{dx}| dx \tag{11}$$

$$\Delta_q = \sqrt{\frac{1}{n}\sum_{i=1}^{n}(\frac{dz_i}{dx})^2} \text{ or } \Delta_q = \sqrt{\frac{1}{l_n}\int_0^{l_n}(\frac{dZ}{dx})^2 dx} \tag{12}$$

Two spatial parameters λ_a and λ_q are the arithmetical mean and the root mean square measure of spatial wavelengths and can be obtained by

$$\lambda_a = 2\pi R_a/\Delta_a \tag{13}$$

$$\lambda_q = 2\pi R_q/\Delta_q \tag{14}$$

A horizontal line at a height z from the mean line can be drawn as shown in figure 4. The distances of intersection between this line and the surface could be measured $(t_i, i = 1,...,m)$. This line could be as high as the R_p value above the mean line and could be as low as the R_v value below the mean line as shown in figure 4. A material

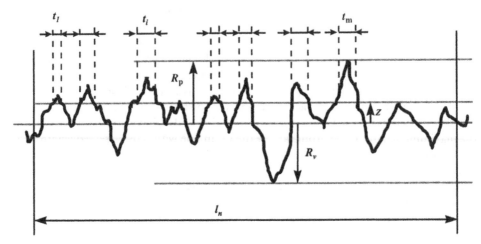

Figure 4.   Illustration of material ratio calculation.

ratio curve $R_{mr}(z)$, also known as the Abbott Firestone curve, for a rough surface can be drawn as shown in figure 5 according to

$$R_{mr}(z) = \frac{\sum_{i=1}^{m} t_i}{l_n} \qquad (15)$$

Three more parameters can be derived from the material ratio curve. A straight line secant to the region at the point of inflection corresponding to a 40% material ratio can be drawn as shown in figure 5. As developed for the German car industry, this straight line is calculated for the central region of the material ratio curve, which gives the minimum mean square deviation in the direction of the surface heights. The straight line intersects the axes of $R_{mr}=0$ and $R_{mr}=1$ at two points which divide the R_t value into three parameters, R_{pk}, R_k and R_{vk}, which are called the reduced peak height, the kernel roughness depth and the reduced trough depth, respectively. However, the process of drawing this straight line is somewhat artificial, depending on how the material ratio curve of the surface height distribution is plotted, as discussed by Whitehouse (1994).

3.   Roughness measurement techniques
Following the work of Harris and Shaw (1988), other instruments have been used to measure surface roughness, in addition to the Surtronic 10, as summarised below.

3.1. *Field measurements*
Most of the surface roughness measurements in the field are done with stylus profilometers. The simplest profilometer is the Surtronic 10. This instrument displays only one surface roughness parameter, either R_a or R_{tm}, in the output. The measuring ranges are 0.1 to 40 μm for R_a and 0.1 to 199.9 μm for R_{tm} with an accuracy of 5% of reading plus 0.1 μm. It has a stylus tip radius of $5 \sim 10$ μm, a cut-off length of 0.8 ±15% mm, a traverse length of 5 mm and an assessed length of 4 mm.

More sophisticated profilometers for field measurements include a Surtronic 3 + by Rank Taylor Hobson and a Surftest by Mitutoyo, Tokyo, Japan. Several filter

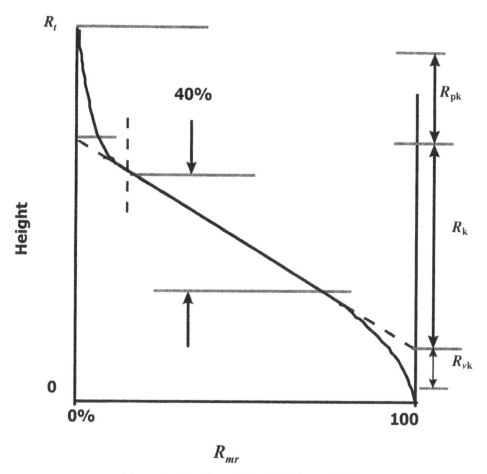

Figure 5. R_{pk}, R_k and R_{vk} (Whitehouse 1994).

types are available such as 2CR, Gaussian or 2CR with phase correction. The 2CR is a recursive filter that complies with International Organization of Standardisation (ISO) standard 3274 (1975). The 2CR with phase correction is also a recursive filter, but with a phase correction. The Gaussian filter is a weighted mean of the profile, where the weights have a Gaussian shape. Since the shape of the Gaussian filter is symmetrical, the resulting filter is phase corrected. Multiple cut-off lengths of 0.25, 0.8 and 2.5 mm, are available. Evaluation lengths can also be selected. There are multiple surface roughness parameter outputs. Vertical resolution for the Surtronic 3+ is 10 nm. An extended processor could be added to the Surtronic 3+ to expand the outputs of surface roughness parameters.

3.2. *Laboratory measurements*
3.2.1. *Stylus profilometers*: Stylus profilometers are the most common devices for surface roughness measurements in laboratories such as profilometers made by Rank Taylor Hobson, Tencor (Mountain View, CA, USA) and Dekdak (Santa Barbara, CA, USA). Usually, several filter types are available. The tip radius of the stylus is within $2 \sim 15$ μm. Waviness measurements are possible

with these types of profilometers. The vertical resolution can vary from 1 Å to 16 nm and the vertical range can vary from 13 μm to 2 mm, depending on the resolution. More cut-off lengths and evaluation lengths are available than with the Surtronic 3+ or Surftest. Some laboratory profilometers, such as some models made by Rank Taylor Hobson and Tencor, can measure surface profile in two dimensions and, thus, generate a three-dimensional surface. However, this type of three-dimensional measurement is very time consuming and the sampling density is usually not as high as with other methods, such as optical devices for three-dimensional measurements, due to the limitation of the control computer.

3.2.1.1. Some limitations of the stylus-type instrument: Stylus-type profilometers have unavoidable limitations in terms of the surface features detectable and difficulties arise when 3D data sets of surfaces are required. The combination of finite tip radius (usually 2 to 15 μm) and included angle prevent the stylus from penetrating fully into deep narrow features of the surface. Special styli, with chisel edges and minimum tip radii as small as 0.1 μm, can be used to examine fine surface details where a conventional stylus would be too blunt, but all stylus methods inevitably produce some 'smoothing' of the profile due to the finite dimensions of the stylus tip. That is, the stylus is large relative to many of the depressions in the surface, so that the stylus rides over them. In some applications, this can lead to significant error. The 3-dimensional approaches provide more detailed information on the micro aspects of surface geometry. This information is a vital source for the investigation of wear mechanisms of shoes and floors.

   A further error can be introduced in examining compliant or smooth surfaces by stylus methods. The load on the stylus, although small, may nevertheless distort or damage the surface so that a non-mechanical contact method, such as an optical interferometry, is recommended for measuring the surface profile (Bhushan *et al.* 1988). In addition to the above limitations, the stylus of the profilometer could bounce on a polymeric material such as shoes due to high resilience of the material. A profilometer may not be suitable for measuring the surface roughness of a macroprofile on soles or heels or a patterned floor surface due to the limit of measurement range.

3.2.2. Other methods for surface geometry measurement: There are wide ranges of instruments for the measurement of surface geometry, which have been used for roughness measurements on shoe or floor surfaces.

3.2.2.1. Optical microscopy (OM):   This is the best known method for observing surfaces with sufficient magnification that it reveals many of the finer features. However, this device suffers from a number of drawbacks. The use of visible light restricts the resolution of the instrument in so far as light is unable to discriminate features that are smaller than 0.25 μm (Culling 1974, Wilson 1985). Furthermore, even at the highest resolution only a very small part of the surface is visible. The optical microscope also lacks depth of focus so that it tends to emphasize the spacing of features rather than their actual height, which is of great interest for observing surface features of both shoes and floors. It is also difficult to quantify surface characteristics using optical microscopy.

3.2.2.2. Electron microscopy (EM): This instrument uses a beam of monoenergetic electrons to produce an image on a fluorescent screen. It gives a much finer resolution of detail than the optical microscope, about 10 Å, which is about 250 times finer than that of the best optical microscope (Thornton 1968, Hren *et al.* 1979). Like the optical microscope, the electron microscope only examines a very small part of the surface, and may therefore misrepresent its general character. This deficiency is overcome in the scanning electron microscope (SEM), which produces a composite picture of a larger surface area by allowing the beam to scan in a controlled sequence (Goldstein *et al.* 1984). Such a system also includes a stereoscopic image of the surface and is therefore an exceedingly useful method of surface examination (Hutchings 1992). However, it is difficult to quantify surface characteristics with this type of microscopy.

3.2.2.3. Interferometry: Light from a common monochromatic source is reflected by a beam-splitting device from the observed surface and from a standard plane reference surface. The combination of these two beams gives rise to a pattern of interference fringes, which are in effect contour lines that indicate the profile of the surface (Chang 1971, Griffiths 1975). Such instruments give a good representation of the surface texture and are self-calibrating since it is known that the vertical distance between adjacent fringes represents one-half of the wavelength of the light used, approximately 0.25 μm. However, this technique only allows one to view a very small, and perhaps unrepresentative, sample of the surface.

3.2.2.4. X-ray diffraction: By studying the diffraction patterns that are obtained when X-rays pass through the surface material at low angles of incidence one is able to ascertain the crystal structure of surfaces (Cullity 1967, Poole 1967). This technique is of interest for the determination of crystal and grain orientation, and surface deformation. The necessary electrons can be readily adapted to most commercial scanning electron microscopes. However, this instrument also cannot quantify surface characteristics.

3.2.2.5. Laser scanning confocal microscope (LSCM): This instrument is a new type of light microscope which has the potential to revolutionise surface analysis. The instrument is similar to conventional confocal microscopes, but it produces better lateral resolutions (Wilson and Sheppard 1984, Wilke 1985, Wilson 1985, van der Voort *et al.* 1987, White and Amos 1987, Wilson and Carlini 1988, Hook and Charles 1989). The distinct advantage of this microscope is its ability to generate rapidly non-destructive optical sections in thick opaque specimens such as shoe and floor surfaces. Another advantage is that the samples need not be specially prepared or mounted as long as they are able to sit on the stage of a conventional microscope. However, a main drawback of this device is that it is not easy to use and requires more experience to become skilful in the operation than other instruments.

4. A summary of the research on floor surface measurements

Following the publication by Harris and Shaw (1988), Stevenson *et al.* (1989) used a pendulum dynamic friction testing machine with a set-up to simulate a heel strike to measure the slip resistance between shoes and floors under five different surface conditions. Steel and concrete floor surfaces with the variation of R_a were used with three shoe heel materials: PVC, nitrile rubber, and leather. They reported that there was a great increase in slip resistance on increasing the roughness from 8 to 28 μm for the concrete surfaces under oily conditions and an increase of the roughness beyond 28 μm only slightly increased the slip resistance. For the PVC and nitrile rubber heels, the increase in COF was from near zero to approximately 0.5. With the leather heel, an increase of the roughness mentioned above did not raise the slip resistance as much as the other two heel materials. For the steel floors with the nitrile heel, very low slip resistance was observed in the oily condition. However, slip resistance was improved to an acceptable level in the wet and soapy conditions. They suggested that the sharpness of the asperities on the surface and the average height are likely to be important.

Grönqvist *et al.* (1990) used another dynamic apparatus to simulate a slip for measuring the slip resistance of several contaminated floor materials. They reported that the Pearson's product-moment and the Spearman rank correlation coefficients between the measured friction and R_a of floor surface roughness were 0.87 and 0.86, respectively, with $p < 0.001$.

Lloyd *et al.* (1992) used the same apparatus as Stevenson *et al.* (1989) to measure slip resistance of 17 different shoes and 15 floor surfaces under four surface conditions. They reported that a linear relationship was found to exist between a roughness index (RI), which contained R_q, λ_q, and R_{sk}, and slip resistance. The multiple correlation coefficient between friction and RI was 0.983 ($p < 0.001$).

Chang (1998) altered the surface roughness on quarry tiles by a sand blasting process and used a pin-on-disk tester to measure dynamic friction with cooking oil as a contaminant. He reported that the correlation coefficient between dynamic friction and R_{pm} of floor surfaces was 0.97 for the averaged surface parameters generated from eight roughness measurements. He also reported that the correlation coefficient between dynamic friction and Δ_a was 0.89 for the surface parameters generated from a roughness measurement.

Chang (1999) used five slipmeters to measure the slip resistance of quarry tiles, which were sand blasted to different degrees of roughness. He reported that the correlation coefficients between the surface parameter R_{pk} and the slip resistance measured under wet conditions with the Ergodyne, also known as the English XL, and the Brungraber Mark II were 0.85 and 0.63, respectively.

Chang (2001) further expanded a previous study (Chang 1998) to two shoe testing materials, Neolite and Four S rubber, on porcelain tiles with contaminants of different viscosity using various mixtures of water and glycerol. The correlation between surface parameter and the measured friction at glycerol contents of 70 and 85% was higher than that at 30 and 99% glycerol contents. Among 22 surface parameters evaluated, Δ_a had the strongest correlation with friction for most of the test conditions with Neolite. For Four S rubber, the parameter R_{pk} appeared to have the strongest correlation with the measured friction coefficient at 30% glycerol content. As the glycerol content was increased to 70 and 85%, the surface parameter Δ_a appeared to correlate strongly with friction for Four S rubber. As the glycerol content was increased further to 99%, the surface parameter R_k had the strongest correlation with the friction of Four S rubber. Chang (2002) extended this study to include vulcanised rubber, a commonly used shoe sole material, and obtained somewhat similar results in a follow-up investigation.

Chang and Matz (2000) and Chang (2000) expanded the two previous studies (Chang 1999, 2001) to investigate the effect of selection of filter types and band widths for surface roughness measurement on the surface roughness parameters and their correlations with friction. Some surface parameters, such as R_{vk} and R_{pk}, were not significantly affected by the filter selection, but they were highly location dependent. Other surface parameters, such as Δ_a and Δ_q, were significantly affected by the filter selection, but they were less location dependent. Moreover, some parameters, such as R_{sk}, were significantly affected by the filter type, while some other parameters, such as Δ_a, were significantly affected by the band width. The results indicated that the filter type 2CR PC could generate surface parameters that had slightly stronger linear correlation coefficients with friction than the ISO 2CR and Gaussian filter types. The surface parameters Δ_a, R_{pm} and R_k were identified to be more suitable for a friction indicator.

5. A summary of the research on shoe surface measurements

The slip resistance of footwear has been studied for many years under various walking conditions. Some of these studies examined the effect of the soling pattern (Perkins and Wilson 1983), and the test method (Strandberg 1985). In addition to the surface roughness on floor surfaces, published studies have confirmed that the surface roughness of shoes substantially influences friction.

Manning *et al.* (1990) measured the surface roughness of solings and floors. They reported that abrasion of rubber solings in stages with increasingly coarse grit gradually raised the value of R_{tm} measured with the Surtronic 10 in parallel with a rise in the COF on wet surfaces, and that both roughness and COF fell during subsequent polishing. The COF was measured by the walking traction method (also known as Rig 3) in which the test-subject walked on the chosen floor surface while pulling against a set of springs until the feet slipped. The springs were attached to a load cell, which measured the horizontal force. It was observed that the COF of footwear increased in parallel with floor roughness. Manning *et al.* (1991) ranked 13 pairs of industrial footwear on Rig 3 in descending order of slip resistance on seven floor surfaces and there was a significant correlation between the R_{tm} of soling and COF. Manning and Jones (1994) measured the slip resistance of a soling compound, microcellular polyurethane (MPU). The surfaces of a series of soling compounds were roughened and then polished and COF measurements were compared on Rig 3. The rough surface of the MPU appeared to be the explanation for its superior slip resistance on wet and oily surfaces. Rowland *et al.* (1996) compared roughness measurements of solings recorded by the Surtronic 10 with photographs taken by a scanning electron microscope (SEM). Dense nitrile rubber solings of worn and new safety boots and a slab of nitrile rubber that was mechanically polished revealed very little surface roughness whereas the MPU had a very rough and irregular surface. Polished samples of microcellular rubbers, which were less slip resistant than the MPU, had a flat surface pitted by bubbles.

In another experiment (Manning *et al.* 1998), five pairs of shoes were soled with slab samples of the same rubber and they were abraded on a sanding machine to give five different degrees of surface roughness on a scale of 4 to 19 μm in R_{tm}. Many readings of COF were taken with Rig 3 on window float glass, smooth vinyl and vinyl tiles coated with wax floor polish under wet conditions. Nearly all of the grip was due to surface roughness of the soling material on these atypically smooth floors, although the surface roughness of the floors also had a significant effect on COF; $p < 0.003$. The authors concluded that the soling roughness was a major factor in determining the COF of this rubber soling material with $p < 10^{-5}$. In a final experiment (Manning and Jones 2001), the COF of 19 soling materials was measured on 19 wet floors and 4 oily floors after various abrasion and polishing treatments of the solings. The profound effect of the roughening of all soles and of floor roughness on the COF were demonstrated for both wet and oily surfaces.

Rowland *et al.* (1996) were early pioneers in examining the microstructures of the shoe heel surfaces as a three-dimensional approach and microscopic approaches were very limited prior to 1996. Only the R_{tm} parameter was considered to represent the roughness characteristics of the shoe surface. The recent studies of Kim (1996a,b, 2000) and Kim and Smith (1999) clearly showed the relationships between slip resistance and extensive range of surface roughness parameters such as the centre line average, maximum height, maximum depth and maximum roughness depth.

6. Identification of the most suitable roughness measurement methods and recommended surface parameters to be measured

6.1. *Field measurements*

The majority of research studies relating shoe and floor roughness to COF were based on measurements of surface roughness taken with the Surtronic 10 and usually the R_{tm} parameter was recorded. These studies have produced numerous statistically significant associations between surface R_{tm} value and COF as measured on walking tests including Rig 3 (Manning and Jones 2001) and the Ramp (Proctor 1993), and by laboratory apparatus (Grönqvist *et al.* 1990 who measured R_a). The UK Health and Safety Executive (HSE 1999) supports the use of the Surtronic 10 and R_{tm} parameter. The Surtronic 10 instrument must therefore be the benchmark instrument for comparison with other methods of measuring roughness but it has limitations and the validity of the R_{tm} parameter is controversial.

At the present time, portable profilometers are the only instruments suitable for measurement of roughness in the field. As indicated previously, the preferred profilometers are the Surtronic 3+ or the Surftest and the preferred surface parameter to be measured is R_{pm} as indicated by UK HSE Food Sheet 22 (HSE 1999) and Chang and Matz (2000). However, field measurements usually have a budget limitation. The Surtronic 10 is one of the most inexpensive profilometers. The output of the Surtronic 10, either R_{tm} or R_a, might not be the parameters with the strongest correlation with friction among those parameters used in the published investigations. In addition, these two surface parameters have significant limitations in representing surface characteristics and are highly location dependent. R_{tm} is a very sensitive indicator of high peaks or deep scratches in a surface (Sherrington and Smith 1987) so that the value of the parameter could be easily changed during the friction process. Also, it should be pointed out that the peak-to-valley values are calculated using data for the whole surface (the assessed length) whereas only the highest points of the surfaces will actually make contact with the shoe heel surfaces of hard soling compounds, for example, the tips of high heel shoes, although softer polymers will drape over the lower peaks (Kim and Smith 2000). However, the correlation between R_{tm} or R_a and the measured friction was statistically significant in the research listed above and, therefore, the Surtronic 10 is recommended for field measurement of surface roughness.

The Surtronic 10 can easily measure flat surfaces such as prepared flat slab solings, but measurement of the roughness of a small area such as the tip of a heel is more difficult. Although the stylus only moves over a length of 4 mm, the base of the Surtronic 10 has to be supported.

6.2. *Laboratory measurements*

The stylus method remains a common way to measure surface profiles in laboratories owing to its earlier establishment and lower cost compared with other methods. However, this category of profilometers has limitations as indicated earlier and non-contact profilometers are preferred. Among non-contact methods listed earlier, the most suitable profilometer is the laser scanning confocal microscope (LSCM) since its outputs are similar to those of both a microscope and a profilometer (Kim and Smith 2000).

Although the surface parameters R_{tm} and R_a were widely used in research, they have significant limitations as indicated earlier. However, it is recommended that R_{tm} and R_a should be included in the measurement outputs until clear evidence suggests otherwise.

Current understanding of the relationships between surface features and friction remains limited. Although the surface roughness parameter that had the strongest correlation with friction depends on the materials and contaminants, the surface parameters R_k, Δ_a, Δ_q, and R_{pm} appeared to have a strong correlation with friction in general (Chang 2002). In addition to surface roughness, friction is also determined by many other factors such as the viscosity and film thickness of the liquid contaminant, sliding velocity at the interface, adhesion and hardness of the two contacting surfaces, hysteresis, and temperature (Moore 1972). The hysteresis loss might be related to the surface slope parameters such as Δ_a and Δ_q since the surface slope is likely to determine the rate of deformation and hence the hysteresis loss in shoe sole material. The surface void volume is likely to determine the area of direct contact between shoe and floor surfaces, so the adhesion might be related to the surface void volume parameters such as R_{pm}. The effect of contaminant film thickness and viscosity might be related to the peak to valley parameters such as R_k since solid to solid contact would occur at the interface if the peak to valley distances are larger than the average film thickness of contaminants. Since the material properties of shoe and floor surfaces and contaminants and the loading conditions at the contact interface determine friction, these factors will determine the surface parameters that have the strongest correlation with the measured friction.

7. Future direction of research and their links to slipperiness measurements

In order to explore the geometric variations, surface topography of both shoes and floor surfaces should be thoroughly examined with suitable surface parameters, and their surface alterations need to be analysed. Three-dimensional images of the surface would provide better information for qualitative and quantitative characterisation of surfaces.

The contribution of surface roughness to slip resistance is linked to the fundamental knowledge of friction and wear behaviours between shoe and floor surfaces. Basic studies on both friction and wear mechanisms would provide vital information in understanding the characteristics of the shoe and floor interface. However, these factors have rarely been considered so far. Simple friction measurements could misrepresent the true slip resistance property. Special caution is required for the measurement of slip resistance under contaminated surface conditions, which are more diverse and complex than the clean and dry surface conditions. Facilitated routine friction measurements in laboratory environments could also oversimplify the intrinsic characteristics of slip resistance between the shoe and floor surface tested.

It is not yet known which method of measuring COF best represents underfoot slip resistance. To date, no comparative studies have shown which method of measuring roughness or which roughness parameter produces the strongest correlation with the COF of a shoe and floor combination during human locomotion. Further research to compare the results of the various methods of measuring roughness will be necessary but cannot be conclusive until the method of measuring COF is agreed. Although there has been considerable progress in understanding the parameters of surface roughness that correlate with the COF values, it is probably true to say that none of the COF measurements on contaminated floors reported to date can be regarded as final objective values for a given floor/soling/contaminant combination.

The literature review showed that most of the publications on surface roughness were related to slipperiness through friction. As friction measurements related to slipperiness remain controversial, the problems with friction measurement can overshadow the contribution of research on surface roughness to slipperiness measurements. Therefore, it is essential for researchers working on surface roughness to find some other measurement methods to assess the slipperiness for complementing current activities. Other methods used to assess slipperiness could include subjective methods and biomechanical methods such as the measurement of slip distance and sliding velocity.

Surface roughness is a critical factor that could significantly affect the slipperiness between footwear and floor interface, but it is not the only factor. Evaluation of other factors such as footwear and floor materials, contaminants, human activities, tread pattern on the footwear and other features on the floor not reflected in the roughness measurements are essential in determining the slipperiness. Further testing is needed to establish a database that could allow users to input other needed information in arriving at the slipperiness indicator.

8. Conclusions

For safe walking without slip and fall accidents, an adequate level of slip resistance between footwear and floor surface is an important issue and it is universally recognized that slippery surfaces are dangerous. The roughened floor surface has a beneficial effect in raising COF values significantly between any shoe heel and floor surface under both dry and contaminated surface conditions. As summarized in the paper, there is convincing evidence that surface roughness on shoe and floor surfaces affect slipperiness significantly. It is suggested that a stylus-type profilometer and a laser scanning confocal microscope are preferred instruments for surface roughness measurements in the field and laboratory, respectively. Furthermore, surface roughness was correlated with slipperiness assessment through friction in most of the published investigations. As the controversy around friction measurement remains, improvement in the methodologies for friction measurement and expansion into other methods of slipperiness measurements are urgently needed. Also, the surface roughness on both shoe and floor surfaces should be measured and combined to arrive at the final assessment of slipperiness. Therefore, surface geometry of both shoe heels and floor surfaces should be monitored routinely to select and maintain best shoe-floor combinations for specific walking environments. This may give a more reliable result for monitoring pedestrian safety than considering the friction measurement alone.

In conclusion, we need a new concept and methodology for the improvement of the slip resistance measurement on pedestrian safety. Rather than simply specifying a minimum value of COF, considerations should be focused on establishing principal approaches for the investigation of complicated slip resistance characteristics. The work should be based on an understanding of the nature of frictional and wear phenomena and surface analysis techniques for characterizing both the surfaces of the shoe and floor and their interactions. Therefore, additional tribological approaches would be a worthwhile way to overcome limitations of existing studies in this research area.

Acknowledgements

The authors would like to thank Margaret Rothwell for her assistance with graphics and manuscript preparation. The authors also thank Robert Brungraber, Chien-Chi Chang, Raoul Grönqvist, Sylvie Leclercq, Lasse Makkonen, Ulrich Mattke, Simon Matz, Rohae Myung, Lennart Strandberg and Steve Thorpe for their thoughtful reviews of the earlier drafts of the manuscript. Helpful discussions on this subject among participants during the symposium were also greatly appreciated.

References

BHUSHAN, B., WYANT, J. C. and MEILING, J. 1988, A new three-dimensional non-contact digital optical profiler, *Wear*, **122**, 301–312.

BOWDEN, F. P. and TABOR, D. 1964, *The Friction and Lubrication of Solids* (Oxford: Clarendon Press).

BRITISH STANDARDS INSTITUTION (BSI) 1988, *Assessment of Surface Texture, Part 1. Methods and Instrumentation*, BS 1134: 1988 (London: British Standards Institution).

CHANG, R. 1971, *Basic Principles of Spectroscopy* (New York: McGraw-Hill), 24.

CHANG, W.-R. 1998, The effect of surface roughness on dynamic friction between Neolite and quarry tile, *Safety Science*, **29**(2), 89–105.

CHANG, W.-R. 1999, The effect of surface roughness on the measurements of slip resistance, *International Journal of Industrial Ergonomics*, **24**, 299–313.

CHANG, W.-R. 2000, The effect of filtering processes on surface roughness parameters and their correlation with the measured friction, Part II: Porcelain tiles, *Safety Science*, **36**(1), 35–47.

CHANG, W.-R. 2001, The effect of surface roughness and contaminant on the dynamic friction of porcelain tile, *Applied Ergonomics*, **32**, 173–184.

CHANG, W.-R. 2002, The effects of surface roughness and contaminants on the dynamic friction between porcelain tile and vulcanized rubber, *Safety Science*, **40**(7 – 8), 577 – 591.

CHANG, W.-R. and MATZ, S. 2000, The effect of filtering processes on surface roughness parameters and their correlation with the measured friction, Part I: Quarry tiles, *Safety Science*, **36**(1), 19–33.

CULLING, C. F. A. 1974, *Modern Microscopy, Elementary Theory and Practice* (London: Butterworth).

CULLITY, B. D. 1967, *Elements of X-ray Diffraction* (Reading, MA: Addison-Wesley).

GOLDSTEIN, J. I., NEWBURY, D. E., ECHLIN, P., JOY, D. C., FIORI, C., LIFSHI, E., HILLMANN, W., KRANZ, O. and ECKOLT, K. 1984, Reliability of roughness measurements using contact stylus instruments with particular reference to results of recent research at the physikalisch technique bundesanstalt, *Wear*, **97**, 27–43.

GRIFFITHS, P. R. 1975, The Michelson interferometer, *Chemical Infrared Fourier Transform Spectroscopy* (New York: John Wiley), 1–14.

GRÖNQVIST, R., ROINE, J., KORHONEN, E. and RAHIKAINEN, A. 1990, Slip resistance versus surface roughness of deck and other underfoot surfaces in ships, *Journal of Occupational Accidents*, **13**, 291–302.

HARRIS, G. W. and SHAW, S. R. 1988, Slip-resistance of floors: User's opinions, Tortus instrument readings and roughness measurement, *Journal of Occupational Accidents,* **9,** 287–298.

HEALTH SAFETY EXECUTIVE (HSE) 1999, Preventing slips in the food and drink industries—A technical update on floor specifications, HSE information sheet: Food Sheet No. 22, HSE Books, Sudbury, Suffolk.

HOOK, G. R. and CHARLES, O. 1989, Confocal scanning fluorescence microscopy: a new method for phagocytosis research, *Journal of Leukocyte Biology,* **45,** 277–282.

HREN, J. J., GOLDSTEIN, J. I. and JOY, D. C. 1979, *Introduction to Analytical Electron Microscopy* (New York: Plenum Press).

HUNTER, R. B. 1930, A method of measuring frictional coefficients of walk-way materials, *National Bureau of Standards Journal of Research,* **3,** 329–348.

HUTCHINGS, I. M. 1992, *Tribology—Friction and Wear of Engineering Materials (Metallurgy & Materials Science Series)* (London: Edward Arnold), 23.

INTERNATIONAL ORGANIZATION OF STANDARDIZATION (ISO) 1975, *Geometrical Product Specifications (GPS)—Surface Texture: Profile Method—Nominal Characteristics of Contact (Stylus) Instruments,* ISO 3274: 1975 (Switzerland: ISO).

INTERNATIONAL ORGANIZATION OF STANDARDIZATION (ISO) 1998, *Geometrical Product Specifications (GPS)—Surface Texture: Profile Method—Terms, Definitions and Surface Texture Parameters,* ISO 4287: 1997 (Switzerland: ISO).

JUNG, K. and REIDIGER, G. 1982, Recent developments regarding the inspection of non-slip floor coverings. Die Berufsgenossenschaft, 1–7, Translation number 12053, Health and Safety Executive, Sudbury, Suffolk.

KENNAWAY, A. 1970, On the reduction of slip of rubber crutch-tips on wet pavements, snow and ice, *Bulletin of Prosthetics Research,* Fall, 130–144.

KIM, I. J. 1996a, Microscopic investigation to analyse the slip resistance of shoes, *Proceedings of the 4th Pan Pacific Conference on Occupational Ergonomics,* November, Taiwan, ROC, 68–73.

KIM, I. J. 1996b, Microscopic observation of shoe heels for pedestrian slip hazard investigation, *Proceedings of the 1st Annual International Conference on Industrial Application and Practice,* December, Houston, TX, 243–250.

KIM, I. J. 2000, Wear progression of shoe heels in dry sliding friction, *Proceedings of the 14th Triennial Congress of the International Ergonomics Association and 44th Annual Meeting of the Human Factors and Ergonomics Society,* August, San Diego, CA.

KIM, I. J. and SMITH, R. 1999, The relationship between wear, surface topography characteristics and coefficient of friction as a means of assessing the slip hazards, *Proceedings of the 2nd Asia-Pacific Conference on Industrial Engineering and Management Systems (APIEMS'99),* October, Ashikaga, Japan, 155–161.

KIM, I. J. and SMITH, R. 2000, Observation of the floor surface topography changes in pedestrian slip resistance measurements, *International Journal of Industrial Ergonomics,* **26,** 581–601.

LLOYD, D. G. and STEVENSON, M. G. 1992, An investigation of floor surface profile characteristics that will reduce the incidence of slips and falls, *Mechanical Engineering Transactions—Institute of Engineers, Australia,* **ME17** (2), 99–105.

MacCURDY, E. 1938, Leonardo da Vinci (1452–1519). Notebooks, translated into English by Edward Jonathan Cape, London.

MANNING, D. P. and JONES, C. 1994, The superior slip-resistance of soling compound T66/103, *Safety Science,* **18,** 45–60.

MANNING, D. P. and JONES, C. 2001, The effect of roughness, floor polish, water, oil and ice on underfoot friction: current safety footwear solings are less slip resistant than microcellular polyurethane, *Applied Ergonomics,* **32,** 185–196.

MANNING, D. P., JONES, C. and BRUCE, M. 1983, Improved slip-resistance on oil from surface roughness of footwear, *Rubber Chemistry and Technology,* **56,** 703–717.

MANNING, D. P., JONES, C. and BRUCE, M. 1990, Proof of shoe slip-resistance by a walking traction test, *Journal of Occupational Accidents,* **12,** 255–270.

MANNING, D. P., JONES, C. and BRUCE, M. 1991, A method of ranking the grip of industrial footwear on water wet, oily and icy surfaces, *Safety Science,* **14,** 1–12.

MANNING, D. P., JONES, C., ROWLAND, F. J. and ROFF, M. 1998, The surface roughness of a rubber soling material determines the coefficient of friction on water lubricated surfaces, *Journal of Safety Research*, **29**, 275–283.

MOORE, D. F. 1972, The friction and lubrication of elastomers, in G. V. Raynor (ed.), *International Series of Monographs on Material Science and Technology*, vol. 9 (Oxford: Pergamon Press).

PERKINS, P. J. and WILSON, M. P. 1983, Slip resistance testing of shoes—new development, *Ergonomics*, **26**, 73–82.

POOLE, C. P., Jr. 1967, *Electron Spin Resonance—A Comprehensive Treatise on Experimental Techniques* (New York: Interscience).

PROCTOR, T. D. 1993, Slipping accidents in Great Britain—an update, *Safety Science*, **16**, 367–377.

RANK TAYLOR HOBSON Ltd 1996, *The Form Talysurf Series 2 Operator's Handbook*, Publication no. K505/9, Leicester, UK.

ROWLAND, F. J., JONES, C. and MANNING, D. P. 1996, Surface roughness of footwear soling materials: relevance to slip-resistance, *Journal of Testing and Evaluation* (JTEVA), **24**, 368–376.

SHERRINGTON, I. and SMITH, E. H. 1987, Parameters for characterizing the surface topography of engineering components, *Proceedings—Institution of Mechanical Engineers*, **201**(C4), 297–306.

SIGLER, A., GEIB, M. N. and BOONE, T. H. 1948, Measurement of the slipperiness of walkway surfaces, Research paper RP1879, *Journal of Research of the National Bureau of Standards*, **40**, 339–346.

STEVENSON, M. G., HOANG, K., BUNTERNGCHIT, Y. and LLOYD, D. G. 1989, Measurement of slip resistance of shoes on floor surfaces, Part 1: Methods, *Journal of Occupational Health Safety—Australia and New Zealand*, **5**, 115–120.

STRANDBERG, L. 1985, The effect of conditions underfoot on falling and overexertion accidents, *Ergonomics*, **28**, 131–147.

THORNTON, P. R. 1968, *Scanning Electron Microscopy* (London: Chapman & Hall).

VAN DER VOORT, H. T. M., BRAKENHOFF, G. T. and JANSSEN, G. C. A. M. 1987, Determination of the 3-dimensional properties of a confocal scanning laser microscope, *Optik*, **78**, 48–53.

WHITE, J. G. and AMOS, W. B. 1987, Confocal microscopy comes of age, *Nature*, **328**, 183–184.

WHITEHOUSE, D. J. 1994, *Handbook of Surface Metrology* (Bristol, UK: Institute of Physics Publishing).

WILKE, V. 1985, Optical scanning microscopy—the laser scan microscope, *Scanning*, **7**, 88–96.

WILSON, T. 1985, Scanning optical microscopy, *Scanning*, **7**, 79–87.

WILSON, T. and CARLINI, A. R. 1988, Three-dimensional imaging in confocal imaging systems with finite sized detectors, *Journal of Microscopy*, **149**, 51–66.

WILSON, T. and SHEPPARD, C. 1984, *Theory and Practice of Scanning Optical Microscope* (London: Academic Press).

CHAPTER 6

The role of friction in the measurement of slipperiness, Part 1: Friction mechanisms and definition of test conditions

WEN-RUEY CHANG†*, RAOUL GRÖNQVIST⁺, SYLVIE LECLERCQ‡, ROHAE MYUNG§,
LASSE MAKKONEN¶, LENNART STRANDBERG††, ROBERT J. BRUNGRABER‡‡,
ULRICH MATTKE§§ and STEVE C. THORPE¶¶

†Liberty Mutual Research Center for Safety and Health, 71 Frankland Road,
Hopkinton, MA 01748, USA

⁺Finnish Institute of Occupational Health, Department of Physics, FIN-00250
Helsinki, Finland

‡French National Research and Safety Institute (INRS), 54501 Vandoeuvre
Cedex, France

§Department of Industrial Systems and Information Engineering, Korea
University, Seongbuk-Gu, Seoul, Korea

¶VTT Building and Transport, Technical Research Centre of Finland, 02044
VTT, Finland

††ITN (Department of Science & Technology), Linköping University, SE-601 74
Norrköping, Sweden

‡‡Department of Civil Engineering, Bucknell University, Lewisburg, PA 17837,
USA

§§Department of Occupational Safety - FB14, University of Wuppertal, D-42097
Wuppertal, Germany

¶¶Health and Safety Laboratory, Sheffield S3 7HQ, UK

Keywords: Friction; Slip resistance; Slipperiness.

Friction has been widely used as a measure of slipperiness. However, controversies around friction measurements remain. The purposes of this paper are to summarise understanding about friction measurement related to slipperiness assessment of shoe and floor interface and to define test conditions based on biomechanical observations. In addition, friction mechanisms at shoe and floor interface on dry, liquid and solid contaminated, and on icy surfaces are discussed. It is concluded that static friction measurement, by the traditional use of a drag-type device, is only suitable for dry and clean surfaces, and dynamic and transition friction methods are needed to properly estimate the potential risk on contaminated surfaces. Furthermore, at least some of the conditions at the shoe/ floor interface during actual slip accidents should be replicated as test conditions for friction measurements, such as sliding speed, contact pressure and normal force build-up rate.

*Author for correspondence. e-mail: Wen.Chang@LibertyMutual.com

1. Introduction

Test methods for the measurement of slipperiness are needed for several purposes. First, they are used to assess slip resistance in various environments in the field and in the laboratory for an improved understanding of the effects on safety of surface roughness and floor contamination. The second purpose is to develop safer products (floorings, footwear and anti-skid devices) through comparative studies of design, construction and material selection. Third, they are used to classify and select anti-slip or slip-resistant products for the purpose of various applications, environments and user groups. Finally, they are used to monitor the effects of maintenance and housekeeping interventions on floor safety, as well as other types of intervention studies at the workplace.

Friction has been widely used as a measure of slipperiness. The general consensus is that surfaces with a lower coefficient of friction (COF) are more slippery than those with a higher COF. Friction has been shown to have a strong correlation with other indicators of slipperiness measurements such as biomechanical, subjective and psycho-physical methods as summarised below.

The biomechanical approach for slip and fall studies has investigated human reaction to slips and falls (Redfern *et al.* 2001). There appear to be two different directional slips during a normal walking step: forward and backward slip during the landing phase and backward slip during the take-off phase (Perkins 1978). Theoretically, it is likely that the forward slip at the landing phase would be the most dangerous due to the body weight being progressively transferred onto the slipping foot. The forward momentum of the body would make it difficult to remove the weight from that foot to regain balance and continued slip would be likely to result in a completely irrecoverable situation.

The relationship between friction coefficient and slip distance at the landing phase was investigated by Perkins (1978), Son (1990), Myung *et al.* (1992) and Myung (1993). Slip distance was shown to be inversely proportional to the static COF for rubber on oily steel (Perkins 1978). It was shown that slip distance was significantly affected by floor conditions (dry and oily) but not walking velocity (Son 1990). Slip distance was shown to be a good measure to differentiate floor slipperiness by an inverse relationship with static and dynamic COF (Myung *et al.* 1992, Myung 1993). Strandberg and Lanshammar (1981) also investigated the dynamics of slipping and concluded that dynamic friction properties seemed to be more important than the static ones for avoiding slips and falls.

Other approaches to measuring slipperiness have included subjective and psycho-physical methods (Grönqvist *et al.* 2001). Subjects are placed in a risk situation and are asked to express safety related to slipping in terms of their sensations or performances. For the subjective method, Tisserand (1969, 1985) used a paired comparison in a subjective safety assessment and drew up a relationship between slipperiness rankings based on COF and subjective assessment. Tisserand (1969) found that the rank order correlation with subjective judgement of slip resistance was negligible for static friction and very high for dynamic friction. Lanshammar and Strandberg (1985) developed a psycho-physical reference method to take into account acceleration and deceleration as well as conditions specific to walking through curves. Subjects were asked to walk as fast as possible without falling in a closed triangular path. The time, T, taken to walk through the experimental path was used to assess the time-based estimate of friction utilisation, μ_T, for the shoe/contaminant/floor combination as follows:

$$\mu_T = K/T^2$$

where K is a function of the course geometry. The model is based on the assumption that the subjects were accelerating and decelerating as much as allowed by the available friction. Hence, the friction estimate obtained from lap time measurements provides an appropriate measure of the practically available friction.

It has been shown in the studies indicated above that friction is a good indicator of slipperiness. However, controversies around friction measurement remain. The arguments appear to concentrate on the validity of different methods for friction measurement. The purposes of this paper are to summarise the current under-standings in friction measurement related to slipperiness measurement, to discuss relevant friction mechanisms and to define test conditions based on the findings. The test conditions defined in the paper will be used to evaluate commonly used slipmeters in Part 2 of the paper.

2. Test conditions for friction measurement from biomechanical data

The ultimate purpose of friction measurement is to prevent slipping accidents, which occur in situations with a huge variability in surface and shoe conditions, and human activities. In order to select proper test conditions, it is necessary to examine biomechanical parameters, knowing that they are influenced by an individual's strategy for coping with the risk of slipping and by the work task performed immediately before slipping.

The dynamics of the forces at the interface between floor and shoe have been studied most often during normal walking in dry, wet or contaminated conditions. Several biomechanical studies on the dynamics of slipping have been published (Perkins 1978, Strandberg and Lanshammar 1981, Redfern and Andres 1984, Soames and Richardson 1985, Hoang *et al.* 1987, Leamon 1988, Leamon and Son 1989, Grönqvist *et al.* 1993, Morach 1993, Redfern and Rhodes 1996, Myung and Smith 1997, Redfern and Di Pasquale 1997). Major observations pointed out by most authors include:

- the dynamics of slipping during different activities;
- a wide variability of parameter values assessed in a laboratory context, even if a specific activity is concerned;
- variations in the gait when normal walking is considered; and
- a general preference for the measurement of a dynamic COF.

To explain the intra- and inter-individual variability, Andres *et al.* (1992) suggested that neurological mechanisms of the humans inside the shoes are not taken into account. Factors such as anticipation of a fall or adaptation of gait, for example, aren't controlled and can modify the dynamics of slipping. This high level of variability lets us imagine a higher variability in the diversity of industrial situations that can be encountered. This variability prevents one from defining precise values for the parameters. Currently only the order of magnitude of parameters can be taken into account.

Biomechanical data alone aren't sufficient to define test conditions for friction measurements. A compromise has to be reached between validity, practicability, reproducibility and variability encountered.

To simulate a slip event, part of the variability of the biomechanical parameters can be taken into account by performing separate tests to simulate different directions of slipping during walking and various work tasks, each test being valid in terms of the situation studied. Reproducibility requires that test conditions be defined precisely and prevents the simulation of some characteristics. For example, Grönqvist *et al.* (1989) obtained a good level of reproducibility of the measurements of slip resistance of shoes on oiled stainless steel if the shoe/surface angle was not varied during measurements.

The choice between static and dynamic friction measurement has been debated for a long time. From a tribological and biomechanical point of view, both would be good parameters. In dry conditions, Myung *et al.* (1992) recommended a static COF. If contaminated conditions are chosen to measure slip resistance of shoes or floors, the measurement of dynamic friction is preferred (Strandberg and Lanshammar 1981, Tisserand 1985, for example).

Strandberg (1983a) suggested that a slipmeter should reproduce the operating variables when slipping during normal straight walking. These variables include contact time, normal force build-up rate, foot angle, contact force application point at the shoe, vertical force, and sliding velocity. He further suggested that the normal force, normal force build-up rate and angle between foot and ground should be of the order of magnitude of 60% of the body weight, 10 kN s^{-1} and 5°, respectively. He also observed that subjects never fell if the peak sliding velocity remained below 0.5 m s^{-1} and suggested that slip resistance testing seemed to be more relevant if the relative sliding speed was kept below 0.5 m s^{-1}. Leamon and Son (1989) found that, during normal straight walking, the angle between foot and ground was approximately 30° except in cases of low and high friction conditions, for which its value was 25°. Myung and Smith (1997) pointed out horizontal heel velocities from 10 to 20 cm s^{-1} on dry floors and from 60 to 140 cm s^{-1} on oily floors, during a load-carrying task. Morach (1993) measured, during normal straight walking, forward foot velocity prior to heel contact of between 0.3 and 2.75 m s^{-1}. Heel velocities during slipping were as high as 2.5 m s^{-1} for a steel floor with oil as a contaminant. Redfern and Rhodes (1996) reported a forward heel velocity of between 0.14 and 0.24 m s^{-1} at heel contact when carrying a box. At that time, heel pitch angle was between 20 and 25°. It decreased to 0° about 100 ms after heel contact.

Contact area and contact pressure at the footwear and floor interface during walking were investigated by Harper *et al.* (1961). Assuming that the shoe sole had a plain surface, the contact area 0.1 s after heel strike was approximately 11 cm^2 and the corresponding contact pressure was approximately 0.41 MPa (Harper *et al.* 1961). The contact pressure from 0 to 0.1 s after a heel strike was extremely high according to their results. Since the contact area measurement was not very precise and the normal contact force was small, there was a large error in the estimate of contact pressure.

3. The role of static, transitional and dynamic friction in slipperiness evaluation

Slipperiness measurement techniques mostly consider either static friction properties or steady-state dynamic friction properties as safety criteria. Static friction properties are measured with the traditional drag-type devices and are limited to the start of a foot slide. Steady-state dynamic friction properties are based on the traditional measurement of resistance to a steady-state motion and are more relevant for an actual steady-state sliding motion between the foot and the walkway. Only a few devices measure transitional friction properties of floor surfaces during a short duration contact time-period between the surfaces (Grönqvist 1997, Grönqvist *et al.* 1999). The transitional friction properties are measured with inclined-strut devices or devices used to simulate a slip from heel strike to steady-state sliding. This transitional test alternative is important for simulating slipping initiated during early heel contact in gait. The moment of heel strike is characterised by collision-type reaction forces built-up within 100 to 200 ms after initial contact with the ground. Assuming that the severity of a foot slide and a subsequent fall incident at a heel strike depends on how quickly and to what extent frictional forces can be built up, then one should perhaps give priority to such a concept for measuring the friction coefficient when contaminants are present at the shoe and floor interface (Grönqvist 1995a, 1999).

Static COF testing may be applicable for the investigation of adhesional friction (true molecular contact) on dry surfaces, but should not be used to assess wet or contaminated surfaces (Strandberg 1983b, 1985). Static friction is affected by the time of stationary contact (dwell time) and by the rate of normal and shear force build-up, including inertial forces at the start of motion (Braun and Roemer 1974, Irvine 1986). In fact, Rabinowicz (1956) disputed the existence of static friction for most materials since any external forces applied to them would produce some creep, i.e. motion. Rabinowicz (1958) also implied that the static COF for short times of stationary contact would equal the dynamic COF.

When a sliding motion starts, transitional COF measurements are valuable for investigating initial frictional squeeze-film processes (drainage and draping) and the evolution of friction over short times (up to 250 ms) of dynamic contact between shoe and floor (Grönqvist 1995a, 1997). Finally, steady-state dynamic COF testing may be used to investigate frictional forces developed after the sliding motion reaches a steady state on contaminated surfaces (Strandberg 1983b).

4. Friction mechanisms

4.1. *Footwear-floor interface*

4.1.1. *Dry surfaces*:   The main contributors to friction of elastomers involved at a dry interface include adhesion, hysteresis and tearing (Tabor 1974, Moore 1975). For elastomeric friction of rubber-like polymers (shoe solings) on dry surfaces, the total COF (μ) due to adhesion and hysteresis depends on contact pressure according to the equation (Moore 1975):

$$\mu = K_1 \left(\frac{sE'}{p^r} + K_2 \left(\frac{p}{E'} \right)^n \right) \tan \delta \qquad (1)$$

In this equation, tan δ is the tangent modulus of the elastomer, defined as the ratio of energy dissipated (E'') to energy stored (E') per cycle. The normal pressure is p, K_1 and K_2 are constants, r is an exponent less than 1 and n is an exponent equal to or greater than 1. The term s is the effective shear strength of the sliding interface. The first term represents the contribution from adhesion, while the second accounts for the contribution from hysteresis. A critical review of hysteresis theories for elastomers is available in the literature (Moore and Geyer 1974).

The main sources of adhesion for polymers are electrostatic and van der Waals forces (Tabor 1974). The contact area between polymers and a hard surface is greater than that predicted by the Hertz theory where no adhesion was assumed (Johnson *et al.* 1971). Increase in contact area due to adhesion is determined by a change in surface energy between two surfaces, asperity radius, combined elastic modulus of two surfaces and normal contact force. When the polymer adheres to another surface, work must be expended in order to produce sliding since the adhesions at the interface must be broken. The formation and rupture of adhesive bonds increase the hysteresis loss of the polymer and, thus, increase the friction. However, experimental results indicate that the strength of adhesion increases with contact duration (dwell time) (Barquins 1982). Therefore, it is essential to control contact duration between shoe and floor surfaces during friction measurements, especially when the surfaces are dry.

Kummer (1966) showed that adhesion and hysteresis of rubber and elastomer friction are manifestations of the same basic viscoelastic energy dissipation mechanism. Adhesion is caused by a dissipative stick-slip process on a molecular level, while hysteresis is an irreversible and delayed response during an elastic contact stress cycle due to damping in materials that leads to energy dissipation, as shown in equation (1). When the stress is removed during travel across a rigid surface, elastomers do not recover completely to their original shape. While adhesion is an energy dissipation process occurring right at the contact interface, hysteresis is an energy dissipation due to deformation that occurs mainly at a small depth below the surface where maximum shear stresses occur.

Recent research has indicated that the shape of the shoe, such as tread pattern, or the geometry, like the bevelled heel, affects friction under dry or contaminated conditions even if the apparent contact surface is identical (Leclercq *et al.* 1995). If the interface between shoe and floor is clean, the basic friction mechanisms are identical regardless of the shoe's shape. However, the shape will affect the boundary conditions and maybe the bulk deformation of the elastomer and, thus, will affect friction indirectly.

The contribution of floor surfaces to friction has not been discussed extensively. Surface roughness on floor surfaces can play a significant role in determining friction at the interface through lubrication, adhesion and hysteresis as indicated by Chang (2002). In addition, the bulk deformation of floor surfaces, especially on resilient floor surfaces or carpet with padding, could potentially generate an interlocking effect similar to the one at the asperity level to increase friction.

4.1.2. *Liquid contaminated surfaces*:   The tribophysics of human slipping, particularly under liquid contaminated conditions, has been discussed by Strandberg (1985), Tisserand (1985), Proctor and Coleman (1988), Grönqvist (1995a, b) and Leclercq *et al.* (1995). Strandberg (1985) pointed out that mechanisms similar to the ones valid for a rolling pneumatic tyre on a wet roadway (Moore 1975) seem to determine walking friction as well. Three major elements at shoe and floor interface include the squeeze-film process and the drainage capability of the shoe-floor contact surface, the draping of the shoe heel and sole about the asperities of the underfoot surface (lubrication, hysteresis) and the true contact between the interacting surfaces (adhesion, hysteresis, tearing). Since draping is time-dependent, slower sliding velocities permit a greater draping effect than higher velocities. Hence, a distinctly higher adhesion is ensured at slow velocities. Hysteretic friction, on the contrary, is small in the low-speed range but becomes greater with increasing sliding velocity and deformation frequency in the interface (Moore 1972).

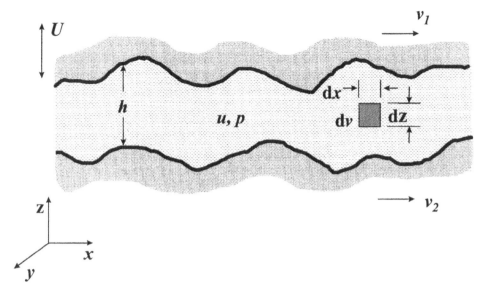

Figure 1.   Schematic of liquid lubricant film governed by the Reynolds' equation.

The generation of hydrodynamic pressure and load support in liquid film shown in figure 1 is a function of wedge, stretch and squeeze terms as shown in the classical Reynolds' equation for liquid in the form of (Moore 1972):

$$\frac{\partial}{\partial x}\left[\frac{h^3}{u}\frac{\partial p}{\partial x}\right] + \frac{\partial}{\partial y}\left[\frac{h^3}{u}\frac{\partial p}{\partial y}\right] = 6(v_1 - v_2)\frac{\partial(h)}{\partial x} + 6h\frac{\partial}{\partial x}(v_1 - v_2) + 12U \qquad (2)$$

where p is the hydrodynamic pressure in the liquid, h is the local instantaneous film thickness, v_1 and v_2 are the horizontal velocities of the upper and lower surfaces, respectively, u is the viscosity of fluid and U is the vertical velocity of the upper surface. The directions of the x and y axes are orthogonal and are parallel to the liquid film surfaces. Moreover, x axis is the direction of velocities v_1 and v_2. The three terms on the right-hand side of the equation represent the contribution of wedge, stretch and squeeze effects, respectively. This equation is valid for smooth surfaces or when the film thickness is much larger than the surface roughness on both surfaces. In the event of boundary lubrication where the normal force is supported by the liquid film and solid to solid contacts, some modifications are needed to account for the effect of surface roughness on both surfaces.

The contribution of the *squeeze term* in equation (2) can be further elaborated, according to Strandberg (1985), as:

$$h^2 = \frac{K \cdot u \cdot A^2}{F_N \cdot t} \qquad (3)$$

where h is the liquid film thickness, K is a shape constant, u is the dynamic viscosity of the fluid, A is the contact area between surfaces, F_N is normal force and t is descending time. When a shoe vertically descends in a liquid film, the film thickness depends on the viscosity of the liquid, the contact area between surfaces, the vertical load and the descending time (Strandberg 1985). The contaminant drainage and draping time for a specified liquid viscosity, normal load and liquid film thickness becomes four times longer when the contact area is doubled. A longer drainage and draping time, on the contrary, may increase the actual risk of slipping because the response time available to prevent a forward slip after heel contact is very short, only a few tenths of a second (Strandberg and Lanshammar 1981). As a consequence, adequate frictional forces between the shoe heel and a smooth contaminated walking surface may not be produced quickly enough to prevent foot sliding and eventual falling.

Leclercq *et al.* (1993) in fact confirmed the importance of the descending time (t) in equation (3) by adjusting the rolling speed of the test wheel of the PFT (Portable Friction Tester) device. When the rolling speed was decreased, descending time increased, allowing film thickness to decrease so that a greater adhesion friction contribution became possible. This contributed to a higher dynamic COF between the test wheel and the substrate.

Figure 2.　PFT prototype device (data as apparatus G in Strandberg 1985). Optional load and wheels provide different forces (F_N) and areas (A) to validate equation (3).

To determine which rolling speeds correspond to real slipping situations, Almshult and Dalén (1983) used a prototype version (called FIDO) of the PFT device as shown in figure 2. By experimenting with viscosity (u), area (A) and load (F_N) they arrived at results supporting Leclercq *et al.* (1993) and the observations mentioned above.

The contribution of the *wedge term* in equation (2) can be further elaborated, according to Proctor and Coleman (1988), as:

$$h^2 = \frac{0.066ul^3v}{F_N} \tag{4}$$

where h and u are film thickness and viscosity of the liquid, respectively, l is length of contact area (a square contact area is assumed), v is sliding velocity and F_N is normal force. Proctor and Coleman (1988) called this the hydrodynamic squeeze-film model. This model is related to the tapered wedge, and it shows that for a specific viscosity, vertical load and slider dimensions, the film thickness varies as the square root of the sliding velocity. The model especially emphasises the importance of reducing the sliding velocity during a slip in order to minimise the contaminant film thickness, and then to obtain good contact and grip between the shoe sole and the floor.

The contribution of the *stretch term* to the pressure generation was not discussed extensively in the literature. However, the stretch term could be critical for the elastomer since it might undergo large deformation during contact with the floor surface.

4.1.3. *Solid contaminated surfaces*: In contrast to liquid contaminants, solid contaminants have received very little attention in research and accident prevention. The mechanisms involved in dry contaminants are still not well understood. Based on some experimental observations, Heshmat *et al.* (1995) hypothesised that the friction mechanisms for dry particulate could include rolling, shearing, normal fracture, elasticity and slip. Kinetic and continuum models were developed for ideal rigid particles. In the kinetic model, it is assumed that all collisions occur in pairs, instead of ternary collisions and other multiple collisions, and elastic collision occurs between particles similar to rarefied gases (Elrod 1988). Different correction factors are introduced to account for different density of particles. However, the kinetic model is not suitable for describing either extremely packed powder masses or the importance of plastic penetration, deformation and interface friction in the development of fractures in the colliding particles. The continuum model basically is a non-Newtonian quasi-hydrodynamic lubrication model derived from some experimental observations (Heshmat *et al.* 1995). Good agreement between the continuum model and experimental results were obtained. More work is needed to expand current understanding of friction mechanisms involved in solid contaminants, especially expanding the current model, with the help of more experimental observations, to particles with irregular shapes and different sizes.

4.2. *Footwear-ice interface*
Bowden and Hughes (1939) showed that very low friction between ice and other materials is due to a water layer formed by frictional heating; however, ice is not always slippery (Petrenko 1994). The ice COF can assume both very small values ($\mu < 0.01$) at high temperatures ($-1°C$) and high velocities (3 m s^{-1}), or very large values ($\mu = 0.67$) at low temperatures ($-40°C$) and low velocities (0.001 m s^{-1}).

The viscoelastic nature of ice friction has been discussed by Moore (1975), but the properties of the interface layer in ice and snow friction are still poorly understood (Makkonen 1994, Kennedy *et al.* 2000).

In general, the properties of ice, such as temperature, structure and hardness as well as the thickness of the water layer, seem to determine friction during a slip to a greater extent than the viscoelastic properties of rubber or polymers (Gnörich and Grosch 1975, Roberts 1981). Low hysteresis and low hardness, which are interrelated in many cases, seem to be necessary properties of rubber to improve friction on ice (Ahagon *et al.* 1988).

4.2.1. *Warm ice*: Assuming that the frictional force, F_μ, in the footwear ice interface is caused mainly by viscous shear in a water layer, the following equation is obtained for COF (Oksanen 1983):

$$\mu = \frac{F_\mu}{F_N} = \frac{\tau A}{F_N} = \frac{u_o v A}{F_N h} \tag{5}$$

where τ is shear stress, A is contact area, u_o is viscosity of water, v is sliding velocity, h is thickness of the water layer and F_N is normal force. F_N/A may be interpreted as the indentation hardness of ice.

Equation (5) is relevant at ice temperatures close to 0°C. The thickness of the water film, h, may be modelled for different sliding configurations and melt water production rates along the lines discussed above (see also Fowler and Bejan 1993). This and the determination of the contact area, A, are inherently difficult, and experiments on wet ice (Jones *et al.* 1994) show a very complex dependence of μ over v. In most cases, however, the COF on wet ice appears to be proportional to the square root of the sliding velocity ($v^{1/2}$). Apparently this is because the film thickness h in equation (5) can be shown to be inversely proportional to $v^{1/2}$ (Oksanen 1983) on a warm ice surface, which is initially dry but melts owing to frictional heating.

4.2.2. *Cold ice*: The problems of using equation (5) can be avoided, when the initial ice temperature is well below 0°C, by considering frictional heating and related thermal balance only. At moderately high temperatures, the interface contact temperature is 0°C due to frictional melting. This allows closure of the system of equations describing the heat balance at the interface and thus makes it possible to consider COF as a dependent variable of the heat balance. Evans *et al.* (1976) and Oksanen and Keinonen (1982) developed semi-quantitative models based on this idea. The resulting COF is proportional to $v^{-1/2}$. This has been partly verified experimentally by Oksanen and Keinonen (1982), Forland and Tatinclaux (1985) and Akkok *et al.* (1987) showing that friction is thermally controlled and the COF essentially depends on the conduction of heat into ice and the slider material (e.g. shoe sole). It is thus possible to control friction on ice by the choice of slider material based on its thermal properties.

At very cold temperatures or very slow velocities, the ice surface is dry as no liquid film is formed. Under these conditions, ice behaves like any ordinary dry surface and has generally a very high COF.

4.2.3. *Other phenomena*: In addition to the friction mechanisms discussed above, asperities of rough sliders will penetrate the ice surface causing plowing. An additional friction mechanism is capillary suction, which increases friction on wet snow (Colbeck 1992). Compaction of snow under the weight of a slider may also squeeze capillary water out of snow thus increasing the thickness of the water layer. Furthermore, electrical forces may affect the COF on dry snow at low temperatures (Petrenko 1994).

As frictional heating does not operate in static friction, one might think that melting of the ice surface does not occur. However, there are two other mechanisms that may cause surface melting of ice below 0°C. First, the ice/water system is anomalous in that its phase change temperature decreases with increasing pressure. Thus, pressure caused by a normal load or an impact (by, for example, a shoe sole) may cause melting at the interface at temperatures very close to 0°C. Consequently, a small change in the normal pressure may cause a rapid and significant reduction in the static COF.

Also, there is a very thin liquid-like layer on an ice surface (in contact with air) at temperatures above approximately −13°C. The origin of this layer is in the imbalance of the molecular forces across the interface (Makkonen 1997). The nature of this imbalance depends on the surface energies and may vary considerably. According to Makkonen's theory (1997), it is possible that at some temperatures the liquid-like layer may disappear in contact with one material but appear in contact with another material. This suggests that the static friction of ice is a complex problem that requires further systematic studies.

4.3. *A flowchart friction model for slipping*
A flowchart friction model for slipping is presented in figure 3 (Grönqvist 1995a). This model takes into account the drainage capability of the shoe-floor contact surface (related to the squeeze-film processes), the draping of the shoe bottom about the asperities of the floor surface (related to deformation and damping), and finally the true contact between the interacting surfaces (related to traction). The squeeze-film processes, occurring between the shoe and walking surface immediately after

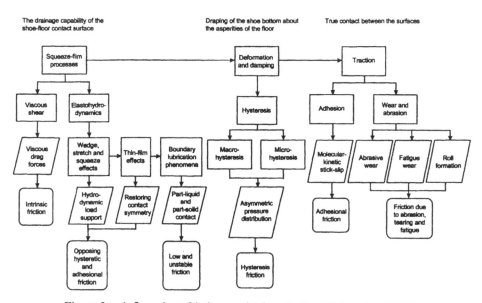

Figure 3.   A flow chart friction model for slipping (Grönqvist 1995a).

first contact at heel touch-down, are critical phenomena affecting pedestrian safety when walking on slippery, contaminated surfaces. One of the key elements in injury prevention is, therefore, the drainage capability of walkways, floorings and footwear solings. If the drainage and draping after heel landing, involving a dynamic loading condition, is too slow due to hydrodynamic load support and elastohydrodynamic effects, then the development of frictional forces will be insufficient. The immediate consequence will be an unstable situation with increased risk of slipping and falling. However, if draping occurs, adequate frictional forces may develop due to deformation and damping (macro- and micro-hysteresis), and may also result in true molecular contact (adhesion and wear) between the interacting surfaces.

5.   Optimal criteria for a slipmeter

For a valid assessment of slipperiness, measurement methodology should meet three basic criteria. First, it should include the measurement of static, transitional, and steady-state dynamic friction properties of the interacting surfaces and contaminants. Second, two different modes of operation, i.e. impact (dynamic loading) and non-impact (static loading), should be included. Third, it should have the flexibility for selecting relevant measurement parameters, such as the normal force build-up time and rate, normal force and pressure, sliding velocity, and contact time of the interacting surfaces prior to and during friction measurement. However, the requirements might be relaxed for devices designed mainly for routine testing in the field.

A dynamic loading test condition, typical for a heel strike in normal walking, has had too little attention in many current slip-testing devices. Only the pendulum strikers, some gait simulators, and some articulated-strut devices produce an impact at the moment of contact, followed immediately by the COF measurement. However, the impact forces produced are mostly poorly defined and do not correspond to normal gait, where the heel touch-down is characterised by collision-type contact forces (Cappozzo 1991).

Test methods applying constant loading, i.e. static or dynamic COF testing without impact, tend to lead to a poorer separation of the interacting surfaces due to lower hydrodynamic pressure generation in the contaminant film (Moore 1972). Hence, they tend to produce higher COF values than the methods applying impact loading when test conditions are identical. Non-impact measurement techniques may underestimate the actual risk of slipping and falling, particularly when wet, oily or greasy conditions are encountered.

Measurement parameters and their ranges should reflect the biomechanics and tribophysics of actual slipping incidents simulating, for instance, the heel strike phase in walking. The normal force build-up rate should be at least 10 kN s^{-1} for the whole-shoe testing devices. The normal pressure at the interface should be between 0.2 and 1.0 MPa, and the sliding velocity should be between zero and 1.0 m s^{-1}. The time of contact prior to and during the COF computation should be between zero and 600 ms.

The validity and reliability of friction-based test methods can be improved if the effects of lubrication and contamination (liquids and solids) at the contact interface between the shoe and walkway could be better understood. The effects of various regimes of lubrication (boundary, mixed and hydrodynamic) on friction measurement and slip resistance are poorly mastered in practice (Grönqvist 1995b). The ratio of viscous to pressure forces in the contaminant film affects the film thickness prior to and during a foot slide (static and dynamic friction, respectively). The contaminant viscosity multiplied by the sliding speed divided by the normal force determines the film thickness and hence the regime of lubrication, which all play key roles for the optimum validity of friction measurements. The choice of test parameters must be optimised separately for various test applications, including simulation of walking on level and inclined surfaces, load carrying, pushing and pulling, etc.

In addition, test methods must be suitable for contaminated floors, particularly liquid contaminants since most accidents occur on wet floors. Tests should use a representative sample of a shoe sole or heel material. Floors should be tested with shoes that are likely to be used on them: gymnasiums with athletic shoes, commercial kitchens with appropriate safety shoes, workshops and constructions sites with work shoes, etc. The tester should be suitable for evaluating and comparing shoes on representative floors as well as evaluating and comparing floors with representative shoes. The tester should be operator insensitive with automated operation and data collection. It should be capable of being compared with readily available and consistent standard materials. This requirement can be met by a combination of low or zero friction shoes and springs or weights to provide well-controlled friction forces. Yet another desirable feature for a laboratory-based slipmeter would be reliable comparability with an equally reliable portable or field-based tester. Some conditions, such as floor care products, that one might wish to evaluate are difficult if not impossible to apply to small samples in exactly the same way in which they are applied in the field. Thus, evaluation of such treatments or products could best be determined with a field-based tester. It would be desirable for such results to be comparable to results on other products, shoes and floors, determined with a laboratory-based tester. A convenient way to meet this requirement would be to develop a slipmeter that is suitable for both laboratory-based and field-based measurements.

6. Conclusions

Friction measurements represent the majority of activities related to prevention in slip and fall accidents. However, current understanding relating friction measurement to slip and fall accidents remains somewhat limited. Moreover, there appear to be regional preferences around the world regarding what devices could best represent human gait involved in slip and fall accidents. These problems further complicate efforts in accident interventions.

Owing to complicated friction phenomena at the shoe and floor interface, it appears that the drag-sled static friction method is only valid for dry and clean surfaces. Dynamic and transition friction methods are needed to properly estimate the potential risk on contaminated surfaces. Furthermore, test conditions should closely resemble the conditions at the shoe and floor interface during actual slip accidents. Based on biomechanical observations during normal walking and friction mechanisms involved at the interface between shoe and floor, test conditions should satisfy some of the conditions, depending on the test methods, as indicated below. The normal force build-up rate should be at least 10 kN s^{-1} for the whole-shoe testing devices. The normal pressure and sliding velocity at the interface should be between 0.2 and 1.0 MPa, and between zero and 1.0 m s^{-1}, respectively. The time of contact prior to and during the COF computation should be between zero and 600 ms. These test conditions will be used as one of the criteria to evaluate commonly used slipmeters both in the field and laboratory settings in Part 2 of this paper (Chang *et al.* 2001).

Acknowledgements

The authors would like to thank Margaret Rothwell for her assistance with manuscript preparation. The authors also thank Yuthachai Bunterngchit, Vincent Ciriello, In-Ju Kim, James Klock and Derek Manning for their thoughtful reviews of the earlier drafts of the manuscript. Helpful discussions on this subject among participants during the symposium were also greatly appreciated. Lasse Makkonen acknowledges the support of the Academy of Finland during the course of this study. This manuscript was completed in part during Dr. Grönqvist's tenure as a researcher at the Liberty Mutual Research Center for Safety and Health.

References

AHAGON, A., KOBAYASHI, T. and MISAWA, M. 1988, Friction on ice, *Rubber Chemistry and Technology*, **61**, 14–35.

AKKOK, M., ETTLES, C. M. and CALABRESE, S. J. 1987, Parameters affecting the kinetic friction of ice, *Journal of Tribology*, **109**, 553–561.

ALMSHULT, B. and DALÉN, B. 1983, Experimental studies of squeeze-film effects relevant to slip-resistance research (in Swedish: Experimentella studier av sjunkförlopp i tunna vätskeskikt för tillämpning inom halkolycksfallsforskning), Masters thesis, Department of Mechanics, Royal Institute of Technology, Stockholm.

ANDRES, R. O., O' CONNOR, D. and ENG, T. 1992, A practical synthesis of biomechanical results to prevent slips and falls in the workplace, in S. Kumar (ed.), *Advances in Industrial Ergonomics and Safety IV* (London: Taylor & Francis), 1001–1006.

BARQUINS, M. 1982, Influence of dwell time on the adherence of elastomers, *Journal of Adhesion*, **14**, 63–82.

BOWDEN, F. and HUGHES, T. 1939, The mechanism of sliding friction on ice and snow, *Proceedings of the Royal Society of London, Series A,* **172**, 280–297.

BRAUN, R. and ROEMER, D. 1974, Influence of waxes on static and dynamic friction, *Soap, Cosmetics, Chemical Specialties*, **50**, 60–72.

CAPPOZZO, A. 1991, The mechanics of human walking, in A. E. Patla (ed.), *Adaptability of Human Gait: Implications for the Control of Locomotion* (Amsterdam: Elsevier), 55–97.

CHANG, W.-R. 2002, The effects of surface roughness and contaminants on the dynamic friction between porcelain tile and vulcanized rubber, *Safety Science*, **40**(7 – 8), 577 – 591.

CHANG, W. R., GRÖNQVIST, R., LECLERCQ, S., BRUNGRABER, R. J., MATTKE, U., STRANDBERG, L., THORPE, S. C., MYUNG, R., MAKKONEN, L. and COURTNEY, T. K. 2001, The role of friction in the measurement of slipperiness, Part 2: Survey of friction measurement devices, Ergonomics, **44**, 1233–1261.

COLBECK, S. C. 1992, A review of the processes that control snow friction, *CRREL Monograph 92-2* (Hanover, NH: US Army Cold Regions Research & Engineering Laboratory).

ELROD, H. G. 1988, Granular flow as a tribological mechanism—a first look, in D. Dowson, C. M. Taylor, M. Godet and D. Berthe (eds), *Interface Dynamics* (Amsterdam: Elsevier), 75–88.

EVANS, D. C. B., NYE, J. F. and CHEESMAN, K. J. 1976, The kinetic friction of ice, *Proceedings of the Royal Society of London, Series A*, **347**, 493–512.

FORLAND, K. A. and TATINCLAUX, J.-C. P. 1985, Kinetic friction coefficient of ice, *CRREL Report 85-6* (Hanover, NH: US Army, Cold Regions Research and Engineering Laboratory).

FOWLER, A. J. and BEJAN, A. 1993, Contact melting during sliding on ice, *International Journal of Heat and Mass Transfer*, **36**, 1171–1179.

GNÖRICH, W. and GROSCH, K. A. 1975, The friction of polymers on ice, *Rubber Chemistry and Technology*, **48**, 527–537.

GRÖNQVIST, R. 1995a, A dynamic method for assessing pedestrian slip resistance, *People and Work*, Research reports 2 (Helsinki: Finnish Institute of Occupational Health), 156.

GRÖNQVIST, R. 1995b, Mechanisms of friction and assessment of slip resistance of new and used footwear soles on contaminated floors, *Ergonomics*, **28**, 224–241.

GRÖNQVIST, R. 1997, On transitional friction measurement and pedestrian slip resistance, in P. Seppälä, T. Luopajärvi, C.-H. Nygård and M. Mattila (eds), *Proceedings of the 13th Triennial Congress of the International Ergonomics Association, IEA '97*, Vol. 3 (Helsinki: Finnish Institute of Occupational Health), 383–385.

GRÖNQVIST, R. 1999, Slips and falls, in S. Kumar (ed.), *Biomechanics in Ergonomics* (London: Taylor & Francis), 351–375.

GRÖNQVIST, R., ABEYSEKERA, J., GARD, G., HSIANG, S. M., LEAMON, T.B., NEWMAN, D.J., GIELO-PERCZAK, K., LOCKHART, T.E. and PAI, Y.-C. 2001, Human-centred approaches in slipperiness measurement, *Ergonomics*, **44**, 1167–1199.

GRÖNQVIST, R., HIRVONEN, M. and TUUSA, A. 1993, Slipperiness of the shoe-floor interface comparison of objective and subjective assessments, *Applied Ergonomics*, **24**, 258–262.

GRÖNQVIST, R., HIRVONEN, M. and TOHV, A. 1999, Evaluation of three portable floor friction testers, *International Journal of Industrial Ergonomics*, **25**, 85–95.

GRÖNQVIST, R., ROINE, J., JÄRVINEN, E. and KORHONEN, E. 1989, An apparatus and a method for determining the slip resistance of shoes and floors by simulation of human foot motions, *Ergonomics*, **32**, 979–995.

HARPER, F. C., WARLOW, W. J. and CLARKE, B. L. 1961, *The Forces Applied to the Floor by the Foot in Walking*, National Building Research Paper 32, DSIR, Building Research Station (London: HMSO).

HESHMAT, H., GODET, M. and BERTHIER, Y. 1995, On the role and mechanism of dry triboparticulate lubrication, *STLE Lubrication Engineering*, **51**, 557–564.

HOANG, K., STEVENSON, M. G., NHIEU, J. and BUNTERNGCHIT, Y. 1987, *Dynamic Friction at Heel Strike Between a Range of Protective Footwear and Non-Slip Floor Surfaces*, Report CSS/1/87 (Kensington, Australia: Centre for Safety Science).

IRVINE, C. H. 1986, Evaluation of the effect of contact-time when measuring floor slip resistance, *Journal of Testing and Evaluation*, **14**(1), 19–22.

JOHNSON, K. L., KENDALL, K. and ROBERTS, A. D. 1971, Surface energy and the contact of elastic solids, *Proceedings of the Royal Society of London A*, **324**, 301–313.

JONES, S. J., KITAGAWA, H., IZUMIYAMA, K. and SHIMODA, H. 1994, Friction on melting ice, *Annals of Glaciology*, **19**, 7–12.

KENNEDY, F. E., SCHULSON, E. M. and JONES, D. E. 2000, The friction coefficient on ice at low sliding velocities, *Philosophical Magazine A*, **80**, 1093–1110.

KUMMER, H. W. 1966, *Unified Theory of Rubber and Tire Friction*, Engineering Research Bulletin B-94 (Pennsylvania: The Pennsylvania State University).

LANSHAMMAR, H. and STRANDBERG, L. 1985, Assessment of friction by speed measurement during walking in a closed path, in D. Winter, R. Norman, R. Wells, K. Hayes and A. Patla (eds), *Biomechanics, IX-B* (Champaign, IL: Human Kinetics Publishers), 72–75.

LEAMON, T. B. 1988, Experimental study of falling, *Proceedings of the Annual Conference of the Human Factors Association of Canada*, Edmonton, Canada, 157–159.

LEAMON, T. B. and SON, D. H. 1989, The natural history of a microslip, in A. Mital (ed.), *Advances in Industrial Ergonomics and Safety I* (London: Taylor & Francis), 633–638.

LECLERCQ, S., TISSERAND, M. and SAULNIER, H. 1993, Quantification of the slip resistance of floor surfaces at industrial sites. Part II. Choice of optimal measurement conditions, *Safety Science*, **17**, 41–55.

LECLERCQ, S., TISSERAND, M. and SAULNIER, H. 1995, Tribological concepts involved in slipping accident analysis, Ergonomics, **38**, 197–208.

MAKKONEN, L. 1994, Application of a new friction theory to ice and snow, *Annals of Glaciology*, **19**, 155–157.

MAKKONEN, L. 1997, Surface melting of ice, *Journal of Physical Chemistry B*, **101**, 6196–6200.

MOORE, D. F. 1972, The friction and lubrication of elastomers, in G. V. Raynor (ed.), *International Series of Monographs on Material Science and Technology*, vol. 9 (Oxford: Pergamon Press).

MOORE, D. F. 1975, *The Friction of Pneumatic Tyres* (Amsterdam: Elsevier).

MOORE, D. F. and GEYER, W. 1974, A review of hysteresis theories for elastomers, *Wear*, **30**, 1–34.

MORACH, B. 1993, Quantifierung des Ausgleitvorganges beim menschlichen Gang unter besonderer Berücksichtigung der Aufsetzphases des Fusses, Fachbereich Sicherheitstechnik der bergischen Universität—Gesamthochschule Wuppertal, Wuppertal.

MYUNG, R. 1993, Floor slipperiness and load carrying effects on the biomechanical study of slips and falls, PhD dissertation, Texas Tech University, Lubbock, TX.

MYUNG, R. and SMITH, J. L. 1997, The effect of load carrying and floor contaminants on slip and fall parameters, *Ergonomics*, **40**, 235–246.

MYUNG, R., SMITH, J. L. and LEAMON, T. B. 1992, Slip distance as an objective criterion to determine the dominant parameter between static and dynamic COFs, *Proceedings of the Human Factors Society 36th Annual Meeting*, October 12–16, 1992, Atlanta, GA, USA, 738–741.

OKSANEN, P. 1983, Friction and adhesion of ice, *Publications 10* (Espoo: Technical Research Centre of Finland).

OKSANEN, P. and KEINONEN, J. 1982, The mechanism of friction on ice, *Wear*, **78**, 315–324.

PERKINS, P. J. 1978, Measurement of slip between the shoe and ground during walking, in C. Anderson and J. Senne (eds), *Walkway Surfaces: Measurement of Slip Resistance*, ASTM Special Technical Publication 649 (Philadelphia: American Society for Testing and Materials), 71–87.

PETRENKO, V. F. 1994, The effect of static electric fields on ice friction, *Journal of Applied Physics*, **76**, 1216–1219.

PROCTOR, T. D. and COLEMAN, V. 1988, Slipping, tripping and falling accidents in Great Britain—present and future, *Journal of Occupational Accidents*, **9**, 269–285.

RABINOWICZ, E. 1956, Stick and slip, *Scientific American*, **194**, 109–118.

RABINOWICZ, E. 1958, The intrinsic variables affecting the stick-slip process, *Journal of Applied Physics*, **29**, 668–675.

REDFERN, M. S. and ANDRES, R. O. 1984, The analysis of dynamic pushing and pulling: required coefficients of friction, *Proceedings of the 1984 International Conference on Occupational Ergonomics*, May 7–9, 1984, Toronto, Ontario, 569–572.

REDFERN, M. S., CHAM, R., GIELO-PERCZAK, K., GRÖNQVIST, R., HIRVONEN, M., LANSHAMMAR, H., MARPET, M. I., PAI, Y.-C. and POWERS, C. M. 2001, Biomechanics of slips, *Ergonomics*, **44**, 1138–1166.

REDFERN, M. S. and DIPASQUALE, J. 1997, Biomechanics of descending ramps, *Gait and Posture*, **6**, 119–125.

REDFERN, M. S. and RHODES, T. P. 1996, Fall prevention in industry using slip resistance testing, in A. Bhattacharya and J. D. McGlothlin (eds), *Occupational Ergonomics, Theory and Applications* (New York: Marcel Dekker), 463–476.

ROBERTS, A. D. 1981, Rubber-ice adhesion and friction, *Journal of Adhesion*, **13**, 77–86.

SOAMES, R. W. and RICHARDSON, R. P. S. 1985, Stride length and cadence: their influence on ground reaction force during gait, in D. Winter, R. Norman and R. Wells (eds), *Biomechanics IX-A* (Champain, IL: Human Kinetics Publishers), 406–410.

SON, D. 1990, The effect of postural changes on slip and fall accidents, PhD dissertation, Texas Tech University, Lubbock, TX.

STRANDBERG, L. 1983a, On accident analysis and slip-resistance measurement, *Ergonomics*, **26**, 11–32.

STRANDBERG, L. 1983b, Ergonomics applied to slipping accidents, in T. O. Kvålseth (ed.), *Ergonomics of Workstation Design* (London: Butterworths), 201–208.

STRANDBERG, L. 1985, The effect of conditions underfoot on falling and overexertion accidents, *Ergonomics*, **28**, 131–147.

STRANDBERG, L. and LANSHAMMAR, H. 1981, The dynamics of slipping accidents, *Journal of Occupational Accidents*, **3**, 153–162.

TABOR, D. 1974, Friction, adhesion and boundary lubrication of polymers, in L.-H. Lee (ed.), *Advances in Polymer Friction and Wear*, Polymer Science and Technology, Vol. 5A (New York: Plenum Press), 5–30.

TISSERAND, M. 1969, Critères d'adhérence des semelles de sécurité, *Rapport d'étude INRS*, 23.

TISSERAND, M. 1985, Progress in the prevention of falls caused by slipping, *Ergonomics*, **28**, 1027–1042.

CHAPTER 7

The role of friction in the measurement of slipperiness, Part 2: Survey of friction measurement devices

Wen-Ruey Chang†*, Raoul Grönqvist⁺, Sylvie Leclercq‡, Robert J. Brungraber§, Ulrich Mattke¶, Lennart Strandberg††, Steve C. Thorpe‡‡, Rohae Myung§§, Lasse Makkonen¶¶ and Theodore K. Courtney†

†Liberty Mutual Research Center for Safety and Health, 71 Frankland Road, Hopkinton, MA 01748, USA

⁺Finnish Institute of Occupational Health, Department of Physics, FIN-00250 Helsinki, Finland

‡French National Research and Safety Institute (INRS), 54501 Vandoeuvre Cedex, France

§Department of Civil Engineering, Bucknell University, Lewisburg, PA 17837, USA

¶Department of Occupational Safety—FB14, University of Wuppertal, D-42097 Wuppertal, Germany

††ITN (Department of Science & Technology), Linköping University, SE-601 74 Norrköping, Sweden

‡‡ Health and Safety Laboratory, Sheffield S3 7HQ, UK

§§ Department of Industrial Systems and Information Engineering, Korea University, Seongbuk–Gu, Seoul, Korea

¶¶ VTT Building and Transport, Technical Research Centre of Finland, 02044 VTT, Finland

This paper seeks to address questions related to friction measurement such as how friction is related to human-centred assessment and actual slipping, and how repeatable friction measurements are. Commonly used devices for slipperiness measurement are surveyed and their characteristics compared with suggested test conditions from biomechanical observations summarised in Part 1. The issues of device validity, repeatability, reproducibility and usability are examined from the published literature. Friction assessment using the mechanical measurement devices described appears generally valid and reliable. However, the validity of most devices could be improved by bringing them within the range of human slipping conditions observed in biomechanical studies. Future studies should clearly describe the performance limitations of any device and its results and should consider whether the device conditions reflect these actual human slipping conditions. There is also a need for validation studies of more devices by walking experiments.

Keywords: Friction; Field base; Laboratory base; Slipperiness; Slipmeter.

*Author for correspondence. e-mail: Wen.Chang@LibertyMutual.com

1. Introduction

Friction has been widely used as a measure of slipperiness. However, controversies about relating friction to slipperiness remain. These include the question of which type of friction is most relevant (static, dynamic, transitional), the problems of inter-method and inter-device variability, and the question of biofidelity between friction as a parameter and slipping as an event. These questions have been examined variously in the other chapters in this book (Chapters 3, 4 and 6).

Despite these limitations, friction measurements remain among the most common approaches to slipperiness evaluation. Numerous devices based on friction measurements, also known as 'slipmeters', have been developed to assess the slipperiness between footwear and walking surfaces. While all of these devices approach the slipperiness question using friction, they differ substantially in their measurement characteristics (reflecting the controversies noted above). Different friction measurement methods have been used, including static drag-sled methods, constant velocity dynamic friction methods and impact methods. Contact pressure, contact area and sliding speed at the interface often vary substantially among devices based on a given type of friction. Pendulum devices measure energy loss when the footwear sample slides through a given distance on the floor surface during a pendulum swing. Owing to diverse measurement characteristics, the results generated from different slipmeters can be quite varied.

In Part 1 (Chang *et al.* 2001), the relationship between friction and slipperiness was discussed. Based on the published literature, there is a strong correlation between friction and slipperiness. Biomechanical observations of the shoe and floor interface were summarised in Part 1 with the particular goal of describing the desirable test conditions for friction measurements related to slip and fall injuries. Friction mechanisms for dry, liquid and solid contaminated as well as icy surfaces were discussed in Part 1 to identify critical parameters for testing conditions. Finally, optimal test conditions were defined based on biomechanical observations and friction mechanisms.

The present paper now addresses the questions related to friction *measurement devices*. As noted, there are significant questions concerning these devices. What relation do they have to human-centred assessment and actual slipping? How repeatable are they? Part 2 surveys commonly-used devices for slipperiness measurement and compares their characteristics with the test conditions defined in Part 1. Evaluation of these devices covers the validity, repeatability, reproducibility, and usability results published in the literature.

2. A survey of devices

2.1. *Field based methods*

The evaluation of floors, as they exist in actual service, can only be conducted under field conditions with standardised shoes and contaminants where possible. Field measurements are typically made in order to:

- evaluate the effect of real use or fouling on the evolution of slip resistance at the same spots in a worksite;
- assess slip resistance at different spots in order to identify the contribution of different factors to slipperiness (Ballance *et al.* 1985, Leclercq *et al.* 1997) and

assess the slipperiness uniformity for identifying areas with a higher risk of fall injuries due to sudden changes in slipperiness (Pater 1985);

• evaluate the slip resistance of a new floor surface. In the case of floors cast or installed on site, the state of the floor surface can be different from that tested in the laboratory even if the two have an identical commercial reference. In that case, measurement is carried out on site, under the same conditions as in the laboratory, usually on a new floor covered with a reference contaminant.

The test conditions of the devices summarised in this section are listed in table 1. These types of devices are portable but can also be used in laboratory studies.

2.1.1. *Static devices/methods*

2.1.1.1. *Horizontal Pull Slipmeter (HPS)*: The HPS (American Society for Testing and Materials [ASTM] F609-96 2001) shown in figure 1 is based on the static drag-sled. The HPS uses a motor to apply the drag force. The contact pressure between the shoe sample and the floor surface is approximately 70.2 kPa. The HPS is intended for use on dry surfaces only.

2.1.1.2. *ASTM C1028-96*: The ASTM C1028-96 (2001), also based on the static drag-sled, uses a dynamometer pull meter to apply a horizontal force by hand. The contact pressure between the shoe sample and floor surface in this case is approximately 38.3 kPa. In contrast to the HPS, the ASTM C1028-96 is

Table 1. Test conditions of field-based friction devices.

Device	Average normal force (N)	Contact area‡ (cm²)	Contact pressure (kPa)	Horizontal sliding velocity (m s⁻¹)	Contact duration for COF computation	Mass (kg)	Dimension (LxWxH in cm)
HPS	2.7	3.8	71	–	–	2.8	26 × 8 × 7
PAST	52	58	9	–	–	14.1	50 × 20 × 50
PFT	112	2.8	400	0.3* (0 – 1)**	0.5 – 3.5 s	42	60 × 40 × 55
AFPV	360* (124, 196, 360)**	16	225* (77, 123, 225)**	0.83	–	–	230 × 72 × 55
Tortus	2	0.6	30	0.017	user specified	6.6	42 × 24 × 10
FSC2000	24	2.5	100	0.2	user specified	7.5	29 × 18 × 11
GMG100	93	11.7	80	0.2 – 0.3	–	9.5	20 × 17 × 19
Shuster	40	26	15	–	user specified	4.1	16 × 10 × 3
BPST	22	2.2	100	2.8	–	13.3	63 × 50 × 70
PIAST	240†	58	41†	–	–	9.5	38 × 20 × 40
VIT	37†	7.9†	47†	–	–	2	30 × 14 × 25
PSM	200* (50 – 500)**	5* (1 – 5)**	400* (100 – 5000)**	0.3* (0 – 0.06)**	any (10 – 5000 ms)	16	42 × 23 × 20

* Typical value; ** adjustable range; † generated with vertical impact; ‡ plain surface. HPS: Horizontal Pull Slipmeter; PAST: Portable Articulated Strut Tribometer; PFT: Portable Friction Tester; AFPV: Low Velocity Skidmeter; FSC 2000: Floor Slide Control 2000; GMG100: GleitMessGerät 100; BPST: British Portable Skid Tester; PIAST: Portable Inclineable Articulated Strut Tribometer; VIT: Variable Incidence Tribometer; PSM: Portable Slipmeter.

recommended for use on both dry and wet surfaces. However, ASTM C1028-96, like other drag-sled approaches, most likely overestimates the coefficient of friction (COF) of wet surfaces.

2.1.1.3. *Portable Articulated Strut Tribometer (PAST)*:   The PAST, also known as the Brungraber Mark I (figure 2), is an articulated-strut slipmeter (ASTM F462-79(1999) 2001, ASTM F1678-96 2001). The COF is determined from the tangent of the angle between the strut and the vertical at which a drastic slip occurs at the interface between footwear sample and floor surface. The contact pressure at the interface is approximately 9.2 kPa. The PAST is intended for measurements on dry surfaces only. An exception to this is for the evaluation of bath tub and shower surfaces in the presence of very soapy water where adhesion has not proven to be a problem so far.

2.1.2. *Steady state dynamic devices*:   All the devices listed in this section measure dynamic COF and can be used on dry and contaminated surfaces. These devices are suitable for soling material and floor testing. Unless explicitly indicated, only one operator is needed to operate each device.

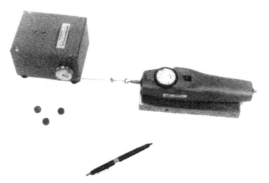

Figure 1.   Horizontal Pull Slipmeter (HPS).

Figure 2.   Portable Articulated Strut Tribometer (PAST).

2.1.2.1. Portable Friction Tester (PFT):   The PFT, as shown in figure 3, has been discussed by Strandberg (1985), Leclercq *et al.* (1993a, b, 1994) and Tisserand *et al.* (1997). This device is pushed at a constant speed by the experimenter on the floor surface. The slip necessary for the measurement of dynamic COF occurs between a braked front wheel (test wheel) and the floor surface. The normal force acting on the front wheel is 112 N. This wheel is covered by a smooth band of elastomer. The apparent contact area between the wheel and the floor surface is 2.8 cm^2 under static conditions. The friction and rotational speeds of the test wheel are a function of the sliding ratio imposed on the front wheel by a speed reducer (0 ~ 100%) and of the speed at which the device is pushed. Using a switch, the operator can determine the times at which the measurement begins and ends. At the end of each measurement, a built-in computer provides estimates of the mean value and standard deviation of the dynamic COF, the measurement distance, the real sliding ratio and the mean pushing speed.

2.1.2.2. Appareil de Frottement à Petite Vitesse—Low Velocity Skidmeter (AFPV):   The AFPV, as shown in figure 4, was discussed by Majcherczyk (1978) and Tisserand *et al.* (1997). The footwear sample is dragged on the floor surface by a guide system at the speed of 0.83 m s^{-1} while normal force is applied through a carriage that slides on the guide rails. The elastomer sample used to measure the floor has slightly bevelled edges and a contact area of approximately 16 cm$^{2.}$

2.1.2.3. Tortus: The Tortus device (Harris and Shaw 1988, Proctor and Coleman 1988, Scheil 1993, Tisserand *et al.* 1997, Grönqvist *et al.* 1999) is pictured in figure 5. This device consists of a four-wheeled trolley driven across the floor surface. To improve its stability on uneven surfaces, one axle is pivoted at the centre giving the device a 3-point support. The footwear sample for COF measurement is attached to a shaft near the centre of the trolley. It moves vertically to accommodate variations in floor level. The drive wheels of the device have a tendency to slip when the surface is oily. Since the pressure and sliding speed are relatively low, the free-floating sample has a tendency to turn on a surface with a raised pattern. This can affect the measurement. The contact area is small, so that the friction measurement can be disrupted by raised patterns. The elastomer sample for COF measurement is a disk with a diameter of 0.9 cm and the contact area is a circle. The measurement is sensitive to oscillation.

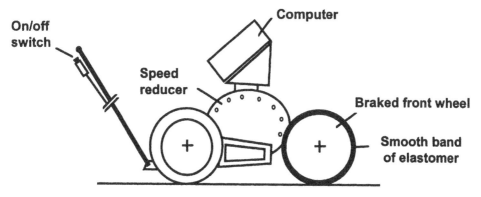

Figure 3.   Portable Friction Tester (PFT) (Leclercq *et al.* 1993b).

Figure 4.   Appareil de Frottement à Petite Vitesse—Low Velocity Skidmeter (AFPV)
(Majcherczyk 1978).

Figure 5.   Tortus device.

2.1.2.4. *Floor Slide Control 2000 (FSC 2000)*:   The FSC 2000 (figure 6) was
discussed by Scheil (1993). The elastomer sample for COF measurement is
hemicylindrical (2.8 × 2.8 cm). The operating principle of this device is similar to
that of the Tortus except that there is an option to lower the shoe sample during the
measurement to include impact loading for transition COF measurement. The drive
wheels of the device also have a tendency to slip when the surface is oily. However, it
has a larger contact area and higher sliding speed than the Tortus and, therefore, can
generate a better lubrication effect.

2.1.2.5. *GleitMessGerät 100 (GMG100)*:   The GMG100, as shown in figure 7, was
originally developed by the German BIA (Berufsgenossenschaftliches Institut für
Arbeitssicherheit) fulfilling the German standard DIN E51131 (1999). This device, in
its final version, drags itself using a steel cable while being anchored by the operator
to a foot plate. The device is driven by an electronically controlled electric motor,
and its motion is stabilised by an internal fly wheel. The device is connected to a

Figure 6.   Floor Slide Control 2000 (FSC 2000).

Figure 7.   GleitMessGerät 100 (GMG100).

notebook computer. Although one person is sufficient to operate this device, in practice it is difficult. Practical tests in the Wuppertal laboratory show significantly lower stick-slip effects than the FSC2000 on most critical materials (Windhövel and Mattke 2002). The fly wheel ensures that the GMG100 moves on floor surfaces smoothly and stably although internal vibrations can be observed.

2.1.2.6. The Schuster: The Schuster device, as shown in figure 8, has been discussed by Scheil (1993) and by Tisserand *et al.* (1997). The operator simply drags the device on floor surfaces to measure friction. The drag speed depends on operator experience. Four thin elastomer samples are used to measure a floor. The dimensions of the elastomer samples are 4 × 1.5 cm. While an experienced operator may use the device alone, two operators are sometimes used (one of them pulls the device and the other records the friction force).

2.1.3. Pendulum swing devices

2.1.3.1. *The Sigler*: The Sigler device, shown in figure 9, measures the energy loss of a pendulum swiping a path of 125–127 mm across a floor surface with a piece of spring-loaded heel material held at an angle with respect to the floor. It was developed in a successful attempt to eliminate the dwell time between shoe and walking surfaces. The dwell time permits the development of adhesion. This device, among the first in its approach, has been succeeded by other more recent devices and is rarely used. Therefore, information about the test conditions of the device was not available for table 1.

2.1.3.2. *The British Portable Skid Tester (BPST)*: This device, as shown in figure 10, embodies many improvements over the Sigler device and is also described by French standard NF P 18-578 (1979), French standard NF P 90-106 (1986), Scheil (1993), Tisserand *et al.* (1997) and Grönqvist *et al.* (1999). The levelling and height-adjusting functions are independent so that set-up time is significantly reduced. The spring that applies the contact force was made as long as the tubular arm of the pendulum and thus more flexible. This results in a more nearly constant contact force between the slider and the walking surface; however, the magnitude of normal force depends on the COF value (Grönqvist *et al.* 1999). Even with these improvements, slow motion photographs taken at the US National Bureau of Standards (NBS) show that the slider bounced across the walking surface (Brungraber 2001).

   The Sigler and the BPST can be used for soling material and floor testing and to measure dry, wet or contaminated surfaces. On surfaces with raised patterns, measurements may be affected if the elastomer hits a bump in the relief at the moment of impact.

2.1.4. Transition friction devices

2.1.4.1. *Portable Inclineable Articulated Strut Tribometer (PIAST)*: The PIAST, also known as the Brungraber Mark II (figure 11), is an inclined-strut slipmeter driven by gravity (ASTM F1677-96 2001). The shoe sample impacts the floor surface with a force and some momentum at an inclined angle from the vertical direction. The COF is obtained from the angle at which a non-slip transitions to a slip. The impact force of the PIAST depends on the angle of impact, the material combination

Figure 8.   Schuster device.

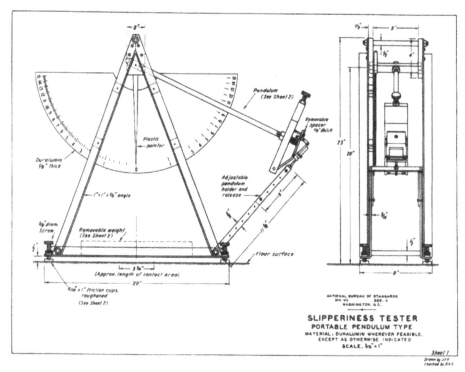

Figure 9. Sigler device.

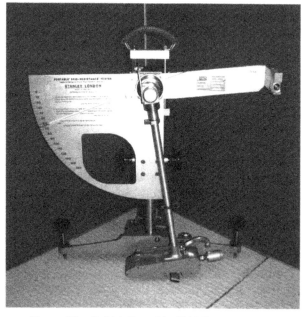

Figure 10. British Portable Skid Tester (BPST).

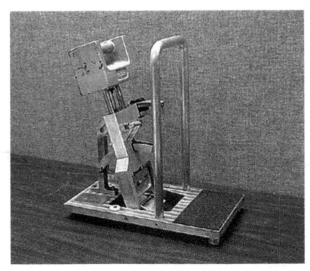

Figure 11.   Portable Inclineable Articulated Strut Tribometer (PIAST) (Chang 2002).

and surface conditions. The peak normal impact force for the PIAST could be as high as approximately 800 N and as low as approximately 300 N at 0° and 26.6° inclinations, respectively, on dry surfaces (Powers *et al.* 1999). Therefore, the mean peak normal contact pressure for the PIAST could be as high as 137.8 kPa and as low as 51.7 kPa at 0° and 26.6° inclinations, respectively. The average normal force for the first impact peak was reported to be in the range of 130 to 240 N corresponding to a mean normal pressure of 22 to 41 kPa (Grönqvist *et al.* 1999).

2.1.4.2. *Variable Incidence Tribometer (VIT)*:   The VIT, known as the English XL (figure 12), is also an inclined-strut slipmeter driven by pneumatic pressure (ASTM F1679-00 2001). The operational principle of the VIT is similar to that of the PIAST. The peak normal impact force for the VIT could be as high as approximately 37 N and as low as approximately 30.6 N at 0° and 35° inclinations, respectively (Powers *et al.* 1999). The initial contact of the shoe sample and the floor is near line contact due to the inclination angle for the VIT; therefore, the normal contact pressure for the VIT immediately after the impact is yet to be established.

2.1.5. *Multi-feature devices*

2.1.5.1. *Portable Slipmeter (PSM)*:   Transitional, dynamic and static friction tests can be performed with the PSM device (figure 13) recently developed by Grönqvist *et al.* (2000). A pneumatic test wheel, comprised of six slider units, is used. The normal force is produced by instantly inflating the pneumatic test wheel, which is being rotated simultaneously at a desired speed just above the floor surface. A triggering device starts the normal force build-up at the moment when the slider approaches the floor surface. The normal force is then quickly developed and constantly maintained over a predetermined time period during each trial. Frictional force data are obtained with a built-in inductive torque transducer, a laptop computer and measurement software. Short duration (less than 200 ms) friction transients (transitional COF) are estimated at the slider/floor interface. Static COF is estimated by rapidly increasing the normal and, within 1s, the shear forces until motion occurs.

Figure 12. Variable Incidence Tribometer (VIT) (Chang 2002).

Figure 13. Portable Slipmeter (PSM).

Alternatively, the static COF can be estimated when a slow speed motion is stopped by stick-slip. Steady-state dynamic COF is estimated over a longer time period of at least 250 ms up to several seconds.

2.2. *Laboratory-based methods*
Evaluation of the slip resistance of shoes, new or used, can best be conducted under laboratory conditions where the walking surface and contaminants can be carefully controlled and thus standardised. Laboratory tests can also be useful to compare new or simulated worn floor surfaces or to study the effect of parameters such as roughness, permeability or simulated wear on COF. The test conditions of the devices summarised in this section are listed in table 2. These types of devices are usually not portable and cannot be used to measure floor surfaces under real operational and environmental conditions.

2.2.1. *Static device*

2.2.1.1. *The James machine*:   The operational principle of the James machine, shown in figure 14, involves the resolution of forces in an inclined strut (James 1944, ASTM D2047-99 2001). A mass of approximately 34 kg is placed on a shaft, which can move in a vertical direction only, for a total mass of approximately 38 kg. Two ends of an articulated strut are separately pinned to the bottom of the shaft and a test shoe material holder. The test shoe material surface, with a dimension of 7.62 × 7.62 cm, is placed on top of the floor surface, which sits on a table that is moveable in a horizontal direction only. The table starts to move at a constant speed with the articulated strut in the vertical position right under the shaft when the test begins, while the angle between the strut and the vertical increases. A slip at the interface between the test shoe material (also called the 'foot') and the floor surface occurs before the vertical shaft reaches the lowest allowable position (It should be noted that the sample thickness determines the effective range of the measurement.) The motions of the vertical shaft and the table are recorded on a chart to derive the COF. The COF is determined from the angle between the strut and the vertical at which a drastic slip happens at the interface between footwear sample and floor surfaces. However, a second approach is used when such a drastic slip does not occur. A non-slip curve is obtained by constraining the position of the test shoe material holder to prevent a slip at the interface. The curve obtained from the measurement is compared with this non-slip curve in the second approach, and COF is then determined when the curve deviates from the non-slip curve, according to ASTM F489-96 (2001).

2.2.2. *Dynamic devices*

2.2.2.1. *French National Research and Safety Institute (INRS) Laboratory Device (LABINRS)*:   This device, shown in figure 15, was discussed by Tisserand (1985), Leclercq *et al.* (1994), French standard XP S 73-012 (1995) and Tisserand *et al.* (1997). The method has been standardised by Association Francaise de Normalisation (AFNOR, French National Standardisation Association) and optimised to measure the dynamic and static COF of safety shoes, soling material, and floors (Leclercq *et al.* 1994). Shoes are tested against a stainless steel floor. For floor tests, 60 × 30 cm flooring samples are tested with a flattened shoe model composed of two semi-cylindrical pieces of smooth elastomer (radius of 17.5 cm and width of 6 cm)

Table 2.   Test conditions of laboratory-based friction devices.

Device/Developer	Contact angle (°)	Normal force (N)	Sliding speed (m s^{-1})
James machine	–	161	–
LABINRS	0 – 20	600* (100 – 1000)**	0.2* (0 – 1.6)**
PSRT	0 – 15	40 – 80	0.01 – 0.2
Stevenson *et al.* (1989)	10	350	0.4* (0 – 0.06)**
Stevenson (1997)	5	490	0.25
STM603	5* (0 – 30)**	400* (up to 750)**	0.1* (0.015 – 0.5)**
Step simulator	0 – 10	350 – 370	0.1 – 0.4
BST 2000	±15	500* (0 – 750)**	0.5
Slip simulator	5* (±25)**	700* (100 – 1500)**	0.4* (0 – 1.0)**

*Typical value; **adjustable range. LABINRS: INRS Laboratory Device; PSRT: Programmable Slip Resistance Tester; STM603: Slip Resistance Tester STM603; BST2000: Boden und Schuhtester 2000.

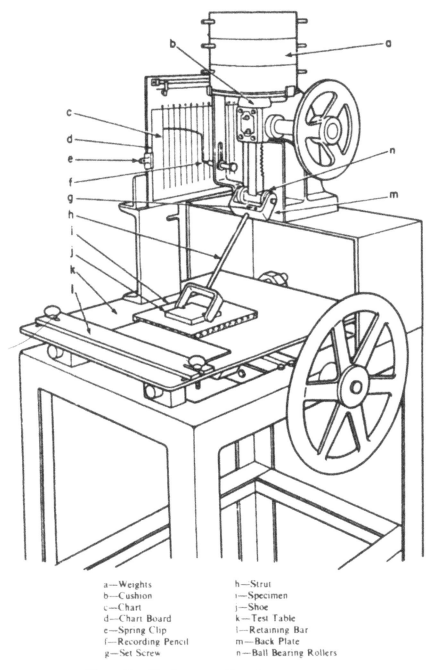

a—Weights   h—Strut
b—Cushion   i—Specimen
c—Chart   j—Shoe
d—Chart Board   k—Test Table
e—Spring Clip   l—Retaining Bar
f—Recording Pencil m—Back Plate
g—Set Screw   n—Ball Bearing Rollers

Figure 14. The James machine (ASTM D2047-99 2001).

with a Shore A hardness of 80. The two pieces of elastomer are 6 × 9 cm, but when they are on the device, the semi-cylindrical shape limits the zone of contact to approximately one-third of the length of each sample. The shoe model withstands a load of 600 N. Either the stainless steel or the surface test sample is moved at a

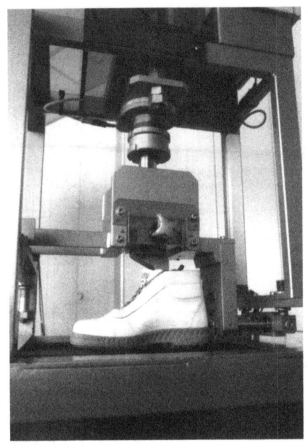

Figure 15.   French National Research and Safety Institute (INRS) Laboratory Device
(LABINRS).

sinusoidally-variable speed. The dynamic COF is measured at the instant when the
speed reaches its maximum value (0.2 m s^{-1}). This method was validated against
subjective rankings (Tisserand 1985). The LABINRS can also measure the COF on
surfaces contaminated with viscous oil or water.

2.2.2.2. *Programmable Slip Resistance Tester (PSRT)*:   Redfern and Bidanda
(1994) developed the PSRT (figure 16) for measuring dynamic COF at the shoe-floor
interface. During a measurement, the shoe performs three movements. First, the
shoe is vertically lowered to the sample floor surface, fully loading the shoe with a
normal force in the range of 40 to 80 N. After approximately 0.5 s, the shoe is moved
across the sample floor. The movement lasts for a total of 5 s and the average
dynamic COF is recorded in a stable, steady-state region which is typically between 3
and 5 s. The shoe to floor angle is between 5° to 15° or it can be flat on the floor. The
sliding velocity is in the range of 0.01 to 0.2 m s^{-1}.

2.2.2.3. *Stevenson* et al. *(1989) device*:   Figure 17 presents the dynamic friction
testing machine developed by Stevenson *et al.* (1989). The test shoe is mounted on a
pendulum. A small hydraulic cylinder connected to the pendulum produces the

normal force between the shoe and floor surfaces. The floor sample is bolted to an instrument that measures the frictional and normal force components. The speed of shoe travel is controlled by another hydraulic cylinder. In a test, the shoe is driven forward by the horizontal cylinder to contact the floor surface at the heel edge, and the normal force is then applied by the vertical cylinder. The angle of the shoe can be adjusted, but typically 10° is chosen. The normal force is set close to 350 N, half of the weight of an average person, which simulates the early heel contact phase. The sliding speed is adjustable up to 0.6 m s^{-1}. However, typically a constant sliding speed of 0.4 m s^{-1} is applied. The dynamic COF is measured from the central section of the recorded trace where the value is stable.

2.2.2.4. *Stevenson (1997) device*: Stevenson (1997) developed a new testing machine (figure 18) where the test shoe is filled with a polyester resin permitting a

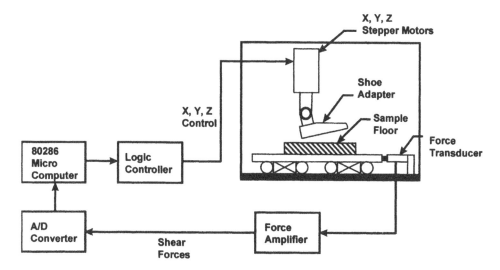

Figure 16. Programmable Slip Resistance Tester (PSRT) (Redfern and Bidanda 1994).

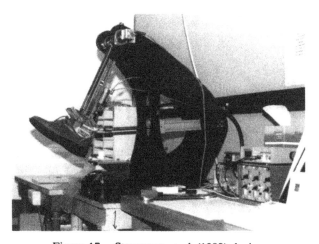

Figure 17. Stevenson *et al.* (1989) device.

relatively rigid connection to the machine. The vertical load of 50 kg is applied partly by a share of mass of the frame itself, but mainly by a set of weights resting on top of the frame. The floor surface sample is screwed to a carriage that can move horizontally. In a test, the pressure of the vertical cylinder is released, allowing it to lower at a rate controlled by a flow valve. When the vertical force is fully applied, the horizontal cylinder is actuated, causing the carriage with the floor sample to move at a controlled speed of 0.25 cm s^{-1} as the dynamic COF is measured. The rate of vertical loading on heel strike is 5 kN s^{-1} and the heel angle at contact is 5° during a typical test run.

2.2.2.5. Slip Resistance Tester STM603:   A new computer-controlled laboratory testing device (figure 19) for measuring friction between shoe and floor surfaces was developed by the Shoe and Allied Trades Research Association (SATRA) using SATRA method TM144 to simulate a slip after a heel strike or before toe-off (Wilson 1990, 2001). The shoe is mounted on a shoemaking form and lowered onto the floor surface. The shoe angle is adjustable up to 30°. A vertical force, adjustable up to 750 N, is applied by means of an air cylinder. After a user-specified delay time of 0 to 5 s, a constant speed motor pulls the surface at a user specified constant velocity of 1.5 to 50 cm s^{-1}. Two load cells are attached to the shoe to measure drag force at the shoe and floor interface up to a total force of 1.2 kN. There are two standard test modes both applying a 400 N vertical force, velocity of 10 cm s^{-1}, and delay time of 0.2 s. These two modes are a forward heel slip with heel at a 5° angle with respect to the floor surface, and a backward shoe-forepart slip with only the centre of the forepart in contact with the floor surface. The dynamic COF is measured 0.3 s after the floor starts to move.

2.2.2.6. Step simulator: The Toegepast-Natuurwetenschappelijk Onderzoek (TNO) friction tester, shown in figure 20, applies a computer controlled step simulator principle (Measurement and Testing Slip Resistance Shoes, Final report 1997). A whole shoe moves while the floor surface remains stationary. The apparatus consists of a test leg module comprised of a mechanical lower leg and foot connected by a rigid ankle. The metal foot can be changed in length and width in order to obtain a good fit between the foot and the shoe. Dynamic COF is calculated from the

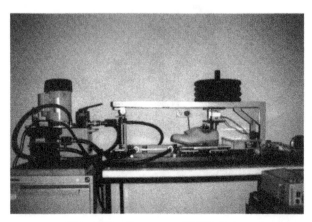

Figure 18.   Stevenson (1997) device.

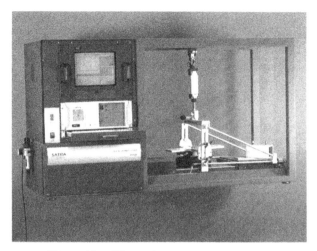

Figure 19.   Slip Resistance Tester STM603.

Figure 20.   Step Simulator.

hydraulically applied normal force and shear force necessary to move the shoe heel or sole relative to the floor surface. Load cells located in the leg module are used to measure the shear and normal forces. Various shoe-floor contact angles (0 to 10°) can be applied. Sliding velocity is also adjustable (0.1–0.4 m s^{-1}). As the normal force is built up, shoe motion in the forward direction begins after a short stationary time period of typically 100 ms or less. Dynamic COF is typically computed between 100 and 450 ms after the sliding begins. The normal force is adjustable between 350 and 750 N.

2.2.2.7. Boden und Schuhtester 2000 (BST 2000):   Skiba *et al.* (1987) developed the floor and shoe tester BST 2000 shown in figure 21. This device consists of a machine frame with drive units for a shuttle table, powered by an electric motor, and a superstructure for the test shoe and a force-measuring element. The shuttle table can be accelerated roughly sinusoidally three times in succession along a distance of 200 mm to a maximum speed of 0.5 m s^{-1}. A typical test cycle repeats this procedure five times in order to obtain an average over the single steps. By means of the superstructure, the shoe can be clamped in at an angle of ±15° from horizontal. Additionally, the shoe is rotatable to 360°. For normal test purposes, an artificial foot is used, but a holder for material samples allows testing for any type of slider. The normal force can range from 0 to 750 N, with a typical value of 500 N. The measurement is automatically controlled by a PC system. Dynamic COF is assessed at the inception of the slide and at a speed of 0.2 m s^{-1}. The values are automatically recorded by the system.

2.2.2.8. Slip Simulator: Grönqvist *et al.* (1989) developed a slip simulator, as shown in figure 22, to simulate a slip following a heel strike in normal walking. An

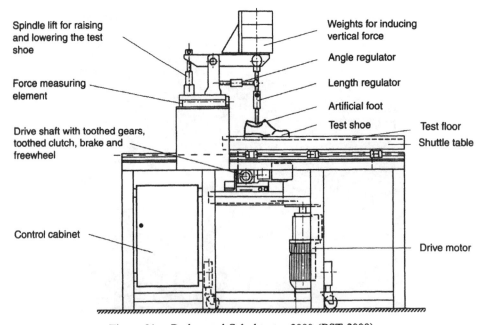

Figure 21.   Boden und Schuhtester 2000 (BST 2000).

artificial foot is used for mounting the shoe to the system. Three cylinders (vertical, horizontal and inclination) powered by a hydraulic device move the shoe against the floor surface. The horizontal cylinder initiates movement in a position slightly above the floor surface and then maintains a selected constant speed, typically 0.4 m s^{-1}, while the vertical cylinder lowers the foot onto the surface at a speed of 0.1 m s^{-1}. Then, the heel of the shoe touches the floor surface at a selected shoe-floor contact angle applied by the inclination cylinder. A force platform measures the frictional and normal forces at the shoe-floor interface. The measurement starts at the moment of heel touch-down, which is normally defined as the instant when the vertical force exceeds 100 N. The time interval of 50 ms for averaging the normal force, friction force, COF and sliding velocity data usually begins 100 ms after heel touch-down. The applied normal force and the sliding velocity are 700 ± 20 N and 0.40 ± 0.02 m s^{-1}, respectively, during a typical test. A shoe-floor contact angle of 5° is frequently used for the heel test. Sole flat, sole backward and sideward sliding tests

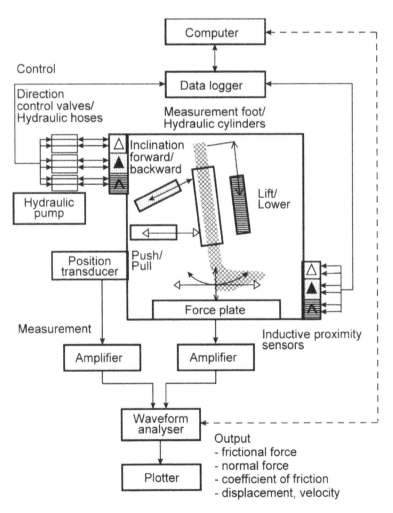

Figure 22. Slip Simulator (Grönqvist *et al.* 1989).

are also possible. The shoe-floor contact angle can be varied in the range ± 25° around an upright position, and the shoe can be rotated 360° in the horizontal plane.

3. Validity

3.1. *Mechanical validity*
The validity of a friction measurement device may be assessed both in terms of its mechanical function (e.g. Does it measure friction forces accurately?) and in terms of its relevance to slip situations (e.g. Does it measure friction under conditions that are reasonably representative of actual footwear/surface interactions?). The first question has been the subject of several studies using force plates or more elaborate laboratory slip measurement devices as reference standards.

The COF values directly obtained with the Portable Inclineable Articulated Strut Tribometer (PIAST) were compared with those calculated from the ground reaction forces generated by the operation of the PIAST on a force plate (Marpet 1996, Marpet and Fleischer 1997, Grönqvist *et al.* 1999, Powers *et al.* 1999). There was good agreement in COF values over different floor surfaces with different surface contaminants for non-slip conditions. The COF measured with this slipmeter was also shown to have a good correlation ($r > 0.954$) with that measured with the slip simulator by Grönqvist *et al.* (1989) although the absolute values of COF from these two devices could be quite different (Grönqvist *et al.* 1999). The COF values directly obtained with the Variable Incidence Tribometer (VIT) on a force plate also were shown to be in good agreement for non-slip conditions (Powers *et al.* 1999).

In a variation of the typical mechanical validity experiment, Chang (1999) investigated the effect of surface roughness on friction measured with five commonly used slipmeters, the PIAST, the VIT, the HPS, the Sigler and the James machine. The VIT and the PIAST coefficient of friction outputs were more strongly correlated with surface roughness parameters under dry and wet surface conditions than the other slipmeters evaluated. The results indicated that at least some friction-based devices can reflect surface roughness conditions.

3.2. *Slipperiness or human-centred validity*
Ideally, a slipmeter would be validated against actual slipping or against slip-related falling events or injuries. However, as Courtney *et al.* (2001) point out, there are still substantial hurdles to be overcome in terms of identifying the events to permit timely assessment. Slipperiness validation studies published to date generally have defined validity as the comparison of the output COF value using a friction measurement device with actual experienced slipping or perceptions of slipperiness (subjective or psychophysical assessments) under controlled walking test conditions. Some have gone beyond this and added biomechanical considerations.

3.2.1. *Published studies of measurement validity:* Tisserand (1985), Leclercq *et al.* (1994) and Tisserand *et al.* (1997) conducted several evaluations involving the rank correlation of results from the INRS Laboratory Device (LABINRS) with the results of paired comparison subjective evaluations. Tisserand (1985) reported a rank correlation of 0.89 in experiments using 10 subjects and 7 different shoe types on oiled stainless steel. Leclercq *et al.* (1994) obtained a rank correlation of 0.91 using

10 subjects wearing successively two different shoe types on 16 oiled floor surfaces. Finally, Tisserand *et al.* (1997) conducted experiments with six devices including the LABINRS using five subjects wearing the same type of safety shoes on 14 oiled floor surfaces. Rank correlation coefficients between the measured COF values of several devices and the results of subjective evaluations based on paired comparisons are shown in column 3 of table 3. The lowest reported correlation coefficient was 0.73 with four of six devices exceeding 0.8.

Rank correlation coefficients between measurements carried out by Harris and Shaw (1988) with the Tortus and results from a subjective rating by users into 1 of 4 divisions for different degrees of slipperiness were 0.78 and 0.38 for dry and wet conditions, respectively. Ten floor surfaces were used for both dry and wet conditions.

Strandberg and Lanshammar (1985) carried out friction measurements on 13 shoe-contaminant-flooring combinations in an interlaboratory comparison of nine apparatuses, i.e. five field devices and four laboratory devices. Two of the selected conditions were dry and clean, one was waxed and the rest were contaminated, since slipping accidents commonly occur on contaminated surfaces. Apparatus-based friction values (AFV) were compared with walking friction data (TFU, time-based friction utilization) measured in walking experiments. Twelve well trained subjects walked a triangular path as fast as they could without slipping and falling. The TFU values were calculated from lap time data according to a model by Lanshammar and Strandberg (1985). The TFU values correlated highly with FFU (force plate-based friction utilization) values over one step during the trials. The correlation coefficients between the AFV and TFU values varied substantially from -0.32 to $+0.92$ (product-moment) and from -0.32 to $+0.98$ (rank order). The highest correlation coefficients were achieved by devices that measured dynamic friction with sufficient velocity, but even the same apparatuses gave totally different AFV values depending on the scaling of the measurement parameters such as sliding velocity and contact area. The correlations between the AFV values of devices measuring static friction were negligible compared to TFU values (cf. Grönqvist *et al.* 2001, about the relationship of TFU and AFV values and actual falling frequencies during walking). In conclusion, devices that were capable of simulating closely the force and motion histories in human gait (e.g. the PFT and the LABINRS) gave the highest correlation as shown in table 3 and, hence, the best validity.

Table 3. Correlation coefficients between friction and walking tests.

Device	Correlation coefficient (Strandberg and Lanshammar 1985)	Rank correlation coefficient (Strandberg and Lanshammar 1985)	Rank correlation coefficient (Tisserand *et al.* 1997)
LABINRS	0.88	0.91	0.80
PFT	0.91	0.96	0.85
Schuster	–	–	0.86
BPST	0.78	0.83	0.74
AFPV	–	–	0.73
Tortus	–	–	0.85

LABINRS: INRS Laboratory Device; PFT: Portable Friction Tester; BPST: British Portable Skid Tester; AFPV: Low Velocity Skidmeter.

Jung and Schenk (1990) compared six different apparatus-based test methods for footwear, two sole-section test methods and two walk test methods. Eight shoes in three floor-contaminant conditions were evaluated against two walk tests. The results shown in table 4 indicated clear relationships between the two walking tests (the BIAT inclined ramp test by the Berufsgenossenschaftliches Institut fur Arbeitssicherheit, and the closed triangular path by the Swedish Road and Traffic Research Institute) and the average friction value for the six machine test methods, except for the triangular walking test on the epoxy floor contaminated with glycerol.

Based on the results of these walking experiments, there appears to be a strong positive correlation between slipmeter outputs and walking experiment outcomes. As detection of fall events and injuries improves, it may eventually become possible to assess surface conditions in near-real time at actual incident scenes. Until this can be accomplished however, walking experiments remain the most likely approach to validation of testing devices. Future experiments could benefit from the use of intentional performance paradigms, such as the lap running in Strandberg and Lanshammars' (1985) experiment. These paradigms simulate a normal purposeful activity and make the subjective assessment of slipperiness a part of that activity as it is in normal human endeavours.

3.2.2. Device parameters that may affect validity

3.2.2.1. General parameters: Measurements based on friction using mechanical test devices, whether laboratory or field instruments, should reproduce the relevant biomechanical and tribophysical parameters during actual slipping events such as a heel slide when slipping during normal walking (Chang et al. 2001, Redfern et al. 2001). Based on biomechanical observations during normal walking and friction mechanisms involved at the interface between the shoe and the floor, the following device parameters should be considered, depending on the test methods, as indicated in Part 1 (Chang et al. 2001):

(1) Normal force build-up rate should be at least 10 kN s^{-1} for whole-shoe devices.
(2) Normal contact pressure should be between 200 and 1000 kPa.
(3) Sliding velocity at the interface should be between zero and 1.0 m s^{-1}. (Static tests would tend to have velocities near zero, while dynamic tests would tend to have velocities closer to the upper bound.)
(4) Maximum time of contact prior to and during the COF computation should be 600 ms.

Table 4.   Correlation coefficients between the mean friction values obtained with six testing machines and the friction data from two walking tests (Jung and Schenk 1990).

Floor/contaminant combination	Walk Test 1 BIAT inclined surface	Walk Test 2 Horizontal triangular path
Steel/Water and detergent (Berol 0.5% wt active content)	0.920**	NA
Steel/Glycerol (Viscosity 200 cP)	0.854*	0.895**
Epoxy/Glycerol (Viscosity 200 cP)	0.936**	0.236

One-sided significance level: *95%; **99%.

In comparing the test conditions of field-based devices shown in table 1 with the parameters in the previous paragraph, the contact pressures of most field devices were below the desired range of 200 to 1000 kPa. The sliding velocity of 2.8 m s^{-1} for the British Portable Skid Tester (BPST) was above the suggested range.

In table 2, the James machine's normal force of 161 N combined with its 58 cm^2 contact area results in a contact pressure of 28 kPa, far below the desired contact pressure range of 200 to 1000 kPa. Among the laboratory-based devices with the capability to mount the whole shoe, the normal contact force for the Programmable Slip Resistance Tester (PSRT) and the Stevenson *et al.* (1989) device appear to be low.

3.2.2.2. *Contamination and squeeze-film effect:* Contaminants, such as water, oil or snow, at the shoe-floor interface are frequently a factor in reduced friction during walking. The presence of such contaminants makes the assessment of slipperiness based on friction difficult due to complex tribophysical phenomena influenced by gait kinematics and kinetics, particularly at heel strike (Chang *et al.* 2001). The squeeze-film effect occurring instantaneously after the first contact at heel touch-down is a critical factor in estimating the risk of slipping during walking on a contaminated surface. The drainage capability of the shoe-floor contact surface and the draping of the shoe bottom about the asperities of the floor are key issues. Adequate frictional forces due to hysteresis and adhesion cannot be developed without a quick penetration of the shoe bottom through the contaminant layer on the walking surface.

When contaminants are present on the surface, it is critical to generate a proper squeeze-film effect, especially for the devices to measure transitional effects at the interface. The squeeze-film theory, shown in equation (3) in Part 1, can be rewritten (Moore 1972) to obtain the rate of reduction in film thickness $\frac{dh}{dt}$ as

$$\frac{dh}{dt} = -\frac{h^3 F_N}{K u A^2} \qquad (1)$$

where K is a shape constant, u is the dynamic viscosity of the fluid, A is the contact area between the surfaces, and F_N is a normal force. As the performance of devices is compared, the initial film thicknesses are identical and the rate of reduction in film thickness, $\frac{dh}{dt}$, is proportional to the value of r_s where $r_s = F_N/A^2$. Therefore, the devices with larger r_s values have less squeeze-film effect than those with smaller values. It is not necessary to generate the same contact area and normal force with the device as would be found at the shoe and floor interface during the critical moments, but an identical r_s value for the contact area and normal force is needed to generate an identical squeeze-film effect.

The calculated r_s values for the field-based portable slipmeters assumed that the contact areas were plain surfaces. The r_s values with the contact area and normal contact force, expressed in cm^2 and N, respectively, for the slipmeters with a capability to measure transition COF listed in table 1 are 3.84 for the FSC2000, 8 for the PSM at the typical test condition, and 0.07 for the PIAST and 0.59 for the VIT when the COF is near zero. However, as the angle of impact with the vertical is increased, the normal contact force and the r_s value decrease for the PIAST. For the VIT, both the contact area and normal force decrease as the angle of impact with the vertical is increased. It appears that the squeeze-film effect generated with the PIAST and the PSM represented the maximum and minimum among the field devices. Among laboratory devices, the STM603, the Step Simulator and the Slip Simulator generate adequate normal force with the capability of having non-zero contact angle between shoe and floor surfaces, generating a better squeeze-film effect than the other laboratory devices.

4. Repeatability, reproducibility and usability

There have been numerous published comparative studies of friction measurement devices and methods examining the repeatability, reproducibility and usability of the various measurement techniques. Reproducibility and repeatability are well-defined quantities (International Organization for Standardization [ISO] 5725 1981). Reproducibility represents the measurement variability from tests performed in different laboratories, with different operators and on different equipment. Repeatability, a less stringent criterion, is defined as precision under repeatable conditions that include the same operator, equipment, and sample within a short time interval. Usability includes a subjective and/or objective evaluation for ease of use. However, these terms can be defined differently, so it should be clear in each case how performance is assessed.

4.1. *Repeatability*

Andres and Chaffin (1985) compared five portable floor slipperiness measurement devices in laboratory and in field conditions: the BPST, the Tortus, and the Friction Instrument Developed by Ohlsson (FIDO) (a pre-prototype of the PFT), the HPS and the PAST. They found that repeatability was variable across the different floor surface materials and floor preparations (dry/wet). The FIDO and the BPST were typically well within $\pm 10\%$ of the mean, while the Tortus showed larger variations. In general, the repeatability of all tested devices was clearly decreased on dry floors versus wet or soapy floors. The reason for the poorer repeatability on dry floors was not explained, but the observation was important since measuring dry friction is a common practice. A recent study by Kim and Smith (2000) indicated that wear on shoe surfaces and wear debris at the interface can be a significant factor on dry surfaces contributing to large changes in friction.

Leclercq *et al.* (1994) discussed the repeatability of floor test measurements carried out with the LABINRS. Friction coefficients were measured over a period of 10 days on 27 oiled floor surfaces with five measurements being made each time per surface with the same equipment by the same operator. For 19 of the 27 samples, the measured COF varied around their means by less than 5%. This variation was greater than 10% in only one single case.

In Jung and Schenks' experiment (1990) introduced in section 3.2.1, the authors reported significant inter-relationships between the various machine test methods. For individual test machines, the repeatability of measured COF values at 95% probability was within 0.06, which was considered to be acceptable.

4.2. *Reproducibility*

Chang and Matz (2001) measured COF of 16 commonly used footwear materials using the PIAST and the VIT on three floor surfaces under four different surface conditions. For contaminated surfaces including wet, oily and oily wet surfaces, the COF obtained with the VIT was typically higher than that measured with the PIAST. The correlation coefficients between the COF obtained with these two slipmeters calculated for different surface conditions indicated a strong correlation with statistical significance.

Braun and Brungraber (1978) found good agreement between two quite different COF testers with a strong statistical significance. One tester was a drag type, having rather sophisticated controls and instrumentation, rendering it non-portable and thus only suitable for laboratory work. The other was the readily portable PAST. Nineteen surfaces, using two test materials, were tested. The strong positive correlation demonstrated that quite different instruments, if well designed and carefully used, can yield similar results, thus lending credibility to both testers.

While inter-device repeatability was generally good, Jung and Schenk (1990) obtained inadequate reproducibility (ISO 5725 1981) among six different test machines. The reproducibility of COF with the 95% probability level was 0.10 on the steel floor contaminated with glycerol, 0.11 on the epoxy floor contaminated with glycerol, and 0.14 on the steel floor contaminated with berol, which is a mixture of water and detergent.

The variability encountered in reliability and reproducibility studies may be attributable, in part, to variations in tested surfaces. Dry friction is hard to reproduce consistently for many materials or pairs of materials, as shown in the often rather broad ranges of values that are reported in many handbook tables (cf. *CRC Handbook of Chemistry and Physics,* 1994–5). In addition, Kim and Smith (2000) have raised the issue of variability due to wear and wear debris which may further complicate the issue. These problems may suggest the need for a standardized testing surface. For example, Q-Panels are standardized, precision-manufactured metal surfaces commonly used for the testing of paints and adhesives. Such a precise and repeatable surface could possibly be employed to reduce the variability due to surface variations when comparing slipmeters.

4.3. *Usability*

The pushing speed of the PFT and the sliding ratio of the test wheel can be slightly different from the set values: the sliding ratio imposed on the test wheel is slightly modified by the resistance from the surface being tested and the pushing speed becomes more constant as the operator becomes more experienced. Consequently, Leclercq *et al.* (1993b) noted that slight deviations from the chosen set values do not significantly influence the friction measurements.

In using the PIAST and the VIT, the operators need to determine if a slip happens at the shoe pad and floor interface. The slip criterion, which is the criterion to determine whether a slip happens, plays a critical role in the performance of these devices. Chang (2002) evaluated the effects of slip criterion on the measured COF with these two slipmeters using two extreme slip criteria. The results indicated that the effect of slip criterion on COF could be quite significant for some material combinations and surface conditions. The results indicated that a more consistent slip criterion is needed.

Andres and Chaffin (1985) subjectively evaluated the ease of use of the BPST, the Tortus, the FIDO, the HPS and the PAST. Their results appear in table 5. Most devices were less sensitive to operator variability and were easy to manoeuvre, while only a few had a straightforward set-up, handled surface irregularities well or had easy to read scales.

5. Conclusions

Friction measurement techniques must be valid and reliable before successful slip prevention strategies can be put into practice. In this paper, the characteristics of friction measurement devices typically used for assessing slipperiness were summarised and evaluated. The validity, repeatability, reproducibility and usability of friction measurement devices were summarised. Test conditions of the devices were compared with those suggested from biomechanical observations as given in Part 1 of this paper.

Based on the results published in the literature, friction assessment using the mechanical measurement devices described herein appears to be generally valid and reliable. However, at least some test conditions in most devices could be improved to bring them within the range of human slipping conditions observed in biomechanical studies to improve their validity. Future studies should describe the performance limitations of any device and its results clearly and should consider whether the device conditions reflect these actual human slipping conditions. Improved understanding of the shoe and floor interface through biomechanical experiments and

Table 5. Usability evaluation of several field-based devices (Andres and Chaffin 1985).

| | | | Ratings | | | |
| | | Poor | | | Good | |
Characteristic	Device	1	2	3	4	5
Setting up to use (included installing	HPS			x		
sensors, calibration check, position in	PAST			x		
test area)	BPST		x			
	Tortus			x		
	FIDO		x			
Readability of scales (and graph of	HPS			x		
the PAST)	PAST			x		
	BPST				x	
	Tortus			x		
	FIDO			x		
Not sensitive to irregularity in floor	HPS			x		
surfaces or curvatures	PAST			x		
	BPST	x				
	Tortus		x			
	FIDO					x
Not influenced by operator methods	HPS				x	
while obtaining measurements	PAST					x
	BPST		x			
	Tortus				x	
	FIDO			x		
Ease of moving device from one	HPS				x	
adjacent surface location to another	PAST					x
	BPST				x	
	Tortus				x	
	FIDO				x	
Ease of packing/unpacking and	HPS					x
carrying to a distant location	PAST			x		
	BPST	x				
	Tortus				x	
	FIDO		x			

HPS: Horizontal Pull Slipmeter; PAST: Portable Articulated Strut Tribometer; BPST: British Portable Skid Tester; Tortus: Tortus device; FIDO: Friction Instrument Developed by Ohlsson.

more extensive validation of devices through walking experiments are needed to improve the relationship between device-based friction assessment and the measurement of slipperiness.

Acknowledgements

The authors would like to thank Margaret Rothwell for her assistance with manuscript preparation. The authors also thank Yuthachai Bunterngchit, Raymond McGorry, In-Ju Kim, James Klock and Derek Manning for their thoughtful reviews of the earlier drafts of the manuscript. Thanks, also, go to Mark Redfern, Michael Wilson, Frans van Hulten and Michael Stevenson for providing some of the photographs and drawings. Helpful discussions on this subject among participants during the symposium were also greatly appreciated. Lasse Makkonen acknowledges the support from the Academy of Finland during the course of this study. This manuscript was completed in part during Dr. Grönqvist's tenure as a researcher at the Liberty Mutual Research Center for Safety and Health.

References

ANDRES, R. O. and CHAFFIN, D. B. 1985, Ergonomic analysis of slip-resistance measurement devices, *Ergonomics*, **28**, 1065-1079.

AMERICAN SOCIETY FOR TESTING AND MATERIALS (ASTM) C1028-96 2001, Standard method for determining the static coefficient of friction of ceramic tile and other like surfaces by the Horizontal Dynamometer Pull-Meter Method, *Annual Book of ASTM Standards*, **15.02**, (West Conshohocken, PA: ASTM).

ASTM D2047-99 2001, Standard test method for static coefficient of friction of polish-coated floor surfaces as measured by the James machine, *Annual Book of ASTM Standards*, **15.04** (West Conshohocken, PA: ASTM).

ASTM F462-79 (1999) 2001, Consumer safety specification for slip-resistant bathing facilities, *Annual Book of ASTM Standards*, **15.07** (West Conshohocken, PA: ASTM).

ASTM F489-96 2001, Standard test method for using a James machine, *Annual Book of ASTM Standards*, **15.07** (West Conshohocken, PA: ASTM).

ASTM F609-96 2001, Standard test method for using a Horizontal Pull Slipmeter (HPS), *Annual Book of ASTM Standards*, **15.07** (West Conshohocken, PA: ASTM).

ASTM F1677-96 2001, Standard test method for using a Portable Inclineable Articulated Strut Slip Tester (PIAST), *Annual Book of ASTM Standards*, **15.07** (West Conshohocken, PA: ASTM).

ASTM F1678-96 2001, Standard test method for using a Portable Articulated Strut Slip Tester (PAST), *Annual Book of ASTM Standards*, **15.07** (West Conshohocken, PA: ASTM).

ASTM F1679-00 2001, Standard test method for using a Variable Incidence Tribometer (VIT), *Annual Book of ASTM Standards*, **15.07** (West Conshohocken, PA: ASTM).

BALLANCE, P. E., MORGAN, J. and SENIOR, D. 1985, Operational experience with a portable friction testing device in university buildings, *Ergonomics*, **28**, 1043–1054.

BRAUN, R. and BRUNGRABER, R. J. 1978, A comparison of two slip-resistance testers, in C. Anderson and J. Seene (eds), *Walkway Surfaces: Measurement of Slip Resistance. ASTM STP 649*, (Philadelphia, PA: ASTM), 49–59.

BRUNGRABER, R. J. 2001, Personal communication.

CHANG, W.-R. 1999, The effect of surface roughness on the measurements of slip resistance, *International Journal of Industrial Ergonomics*, **24**, 299–313.

CHANG, W.-R. 2002, The effects of slip criterion and time on friction measurements, *Safety Science*, **40**(7–8), 593–611.

CHANG, W.-R., and MATZ S., 2001, The slip resistance of common footwear materials measured with two slipmeters, *Applied Ergonomics*, **32**, 549–558.

CHANG, W.-R., GRÖNQVIST, R., LECLERCQ, S., MYUNG, R., MAKKONEN, L., STRANDBERG, L., BRUNGRABER, R. J., MATTKE, U. and THORPE, S. C. 2001, The role of friction in the measurement of slipperiness, Part 1: Friction mechanisms and definition of test conditions, *Ergonomics*, **44**, 1217–1232.

COURTNEY, T. K., SOROCK, G. S., MANNING, D. P., COLLINS, J. W. and HOLBEIN-JENNY, M. A. 2001, Occupational slip, trip, and fall-related injuries—can the contribution of slipperiness be isolated?, *Ergonomics,* **44,** 1118–1137.

CRC *Handbook of Chemistry and Physics 1994–5,* 75th edn, (Cleveland, OH: CRC Press), 15–40.

DEUTSCHES INSTITUT FüR NORMUNG (DIN) E51131 1999, *Prüfung von Bodenbelägen— Bestimmung der rutschhemmenden Eigenschaft—Verfahren zur Messung des Gleitrei- bungskoeffizienten,* [Testing of floor coverings—Determination of the anti-slip proper- ties—Measurement of sliding friction coefficient] (Berlin: DIN).

GRÖNQVIST, R., ABEYSEKERA, J., GARD, G., HSIANG, S. M., LEAMON, T. B., NEWMAN, D. J., GIELO-PERCZAK, K., LOCKHART, T.E. and PAI, Y.-C. 2001, Human-centred approaches in slipperiness measurement, *Ergonomics,* **44,** 1167–1199.

GRÖNQVIST, R., CHANG W.-R., HIRVONEN, M., RAJAMäKI, E. and TOHV, A. 2000, Validity and reliability of transitional floor friction tests: the effect of normal load and sliding velocity, *Proceedings of the XIVth Triennial Congress of the International Ergonomics Association and 44th Annual Meeting of the Human Factors and Ergonomics Society,* July 29–August 4, San Diego, CA, **4,** 502–505.

GRÖNQVIST, R., HIRVONEN, M. and TOHV, A. 1999, Evaluation of three portable floor slipperiness testers, *International Journal of Industrial Ergonomics,* **25,** 85–95.

GRÖNQVIST, R., ROINE, J., JäRVINEN, E. and KORHONEN, E. 1989, An apparatus and a method for determining the slip resistance of shoes and floors by simulation of human foot motions, *Ergonomics,* **32,** 979–995.

HARRIS, G. W. and SHAW, S. R. 1988, Slip resistance of floors: users' opinions, Tortus instrument readings and roughness measurement, *Journal of Occupational Accidents,* **9,** 287–298.

INTERNATIONAL ORGANIZATION FOR STANDARDIZATION (ISO) 5725-1981(E) *Precision of Test Methods—Determination of Repeatability and Reproducibility by Inter-Laboratory Tests,* 1st edn, 1981-04-01, (Geneva:ISO).

JAMES, S. V. 1944, What Is a Safe Floor Finish?, *Soap and Sanitary Chemicals,* **20,** 111–115.

JUNG, K. and SCHENK, H. 1990, An international comparison of test methods for determining the slip resistance of shoes, *Journal of Occupational Accidents,* **13,** 271–290.

KIM, I. J. and SMITH, R. 2000, Observation of the floor surface topography changes in pedestrian slip resistance measurements, *International Journal of Industrial Ergonomics,* **26,** 581–601.

LANSHAMMAR, H. and STRANDBERG, L. 1985, Assessment of friction by speed measurement during walking in a closed path, in D. Winter, R. Norman, R. Wells, K. Hayes and A. Patla (eds), *Biomechanics, IX-B* (Champaign, IL: Human Kinetics Publishers), 72–75.

LECLERCQ, S., TISSERAND, M. and SAULNIER, H. 1993a, Quantification of the slip resistance of floor surfaces at industrial sites. Part 1. Implementation of a portable device, *Safety Science,* **17,** 29–39.

LECLERCQ, S., TISSERAND, M. and SAULNIER, H. 1993b, Quantification of the slip resistance of floor surfaces at industrial sites. Part 2. Choice of optimal measurement conditions, *Safety Science,* **17,** 41–55.

LECLERCQ, S., TISSERAND, M. and SAULNIER, H. 1994, Assessment of the slip-resistance of floors in the laboratory and in the field: two complementary methods for two applications, *International Journal of Industrial Ergonomics,* **13,** 297–305.

LECLERCQ, S., TISSERAND, M. and SAULNIER, H. 1997, Analysis of measurements of slip resistance of soiled surfaces on site, *Applied Ergonomics,* **28,** 283–294.

MEASUREMENT AND TESTING SLIP RESISTANCE SHOES 1997, Development of a test method for measuring the slip resistance of protective footwear. Unpublished final report (20 October, 1997) for the Commission of the European Communities, Contract no. MAT1 CT 940059.

MAJCHERCZYK, R. 1978, A different approach to measuring pedestrian friction: the CEBTP skidmeter, in C. Anderson and J. Senne (eds), *Walkway Surfaces: Measurement of Slip Resistance,* ASTM STP 649, (Philadelphia: ASTM), 88–99.

MARPET, M. I. 1996, Comparison of walkway-safety tribometers, *Journal of Testing and Evaluation,* **24,** 245–254.

MARPET, M. I. and FLEISCHER, D. H. 1997, Comparison of walkway-safety tribometers: Part 2, *Journal of Testing and Evaluation,* **25,** 115–126.

MOORE, D. F. 1972, The friction and lubrication of elastomers, in G. V. Raynor, (ed.), *International Series of Monographs on Material Science and Technology, Vol. 9*, (Oxford: Pergamon Press).

NF P 18-578, 1979, *Granulats—Mesure de la rugosité d'une surface à l'aide du pendule de frottement*. [Aggregates—Measurement of the surface rugosity with the friction pendulum], (Paris la défense: AFNOR).

NF P 90-106, 1986, *Sol sportifs—Mesure de la glissance d'une surface à l'aide d'un pendule de frottement*, [Sports grounds—Measurement of the skidding conditions of a surface with a friction pendulum]. (Paris la défense: AFNOR).

PATER, R. 1985, How to reduce falling injuries, *National Safety and Health News*, October, 87–91.

POWERS, C. M., KULIG, K., FLYNN, J. and BRAULT, J. R. 1999, Repeatability and bias of two walkway safety tribometers, *Journal of Testing and Evaluation*, **27**, 368–374.

PROCTOR, T. D. and COLEMAN, V. 1988, Slipping, tripping and falling accidents in Great Britain—present and future, *Journal of Occupational Accidents*, **9**, 269–285.

REDFERN, M. S. and BIDANDA, B. 1994, Slip resistance of the shoe-floor interface under biomechanically-relevant conditions, *Ergonomics*, **37**, 511–524.

REDFERN, M. S., CHAM, R., GIELO-PERCZAK, K., GRÖNQVIST, R., HIRVONEN, M., LANSHAMMAR, H., MARPET, M. I., PAI Y.-C. and POWERS, C. M. 2001, Biomechanics of slips, *Ergonomics*, **44**, 1138–1166.

SCHEIL, M. 1993. Analyse und Verlgleich von instationären Reibzahlmessgeräten. [Analysis and comparison of instationary friction coefficient measurement tools.] Fachbereich Sicherheitstechnik der Bergischen. [Department of Safety Technique]. Doctoral dissertation in German,with English summary, Universität-Gesamthochschule, Wuppertal.

SKIBA, R., KUSCHEFSKI, A. and CZIUK, N. 1987, *Entwicklung eines normgerechten Prüfverfahrens zur Ermittlung der Gleitsicherheit von Schuhsohlen, Forschung Fb 526*. [Development of a standardized method to determine the slip resistance of shoe soles, Research report no. 526] Schriftenreihe der Bundesanstalt für Arbeitsschutz, Dortmund, (Bremerhaven: Wirtschaftsverlag NW).

STEVENSON, M. 1997, Evaluation of the slip resistance of six types of women's safety shoe using a newly developed testing machine, *Journal of Occupational Health and Safety— Australia and New Zealand*, **13**, 175–182.

STEVENSON, M. G., HOANG, K., BUNTERNGCHIT, Y. and LLOYD, D. 1989, Measurement of slip resistance of shoes on floor surfaces. Part 1. Methods, *Journal of Occupational Health and Safety—Australia and New Zealand*, **5**, 115–120.

STRANDBERG, L. 1985, The effect of conditions underfoot on falling and overexertion accidents, *Ergonomics*, **28**, 131–147.

STRANDBERG, L. and LANSHAMMAR, H. 1985, Walking slipperiness compared to data from friction meters, in D. Winter, R. Norman, R. Wells, K. Hayes and A. Patla (eds.), *Biomechanics IX B*, (Champaign, IL: Human Kinetics Publishers), 76–81.

TISSERAND, M. 1985, Progress in the prevention of falls caused by slipping, *Ergonomics*, **28**, 1027–1042.

TISSERAND, M., SAULNIER, H. and LECLERCQ, S. 1997, Comparison of seven test methods for the slip resistance of floors: contributions to developments of standards, in P. Seppälä, T. Luopajärvi, C.-H. Nygård and M. Mattila (eds), *Proceedings of the 13th Triennial Congress of the IEA*, Vol. 3 (Helsinki: Finnish Institute of Occupational Health), 406–408.

WILSON, M. P. 1990, Development of SATRA slip test and tread pattern design guidelines, in B. E. Gray (ed), *Slips, Stumbles, and Falls: Pedestrian Footwear and Surfaces*, ASTM STP 1103 (Philadelphia: ASTM), 113–123.

WILSON, M. P. 2001, Personal communication.

WINDHÖVEL, U. and MATTKE, U. 2002, Neuere Entwicklung im Regelwerk über die Rutschfestigkeit von Schuhen und Fußböden, Sicherheitsingenieur [New developments in the safety requirements for the slip resistance of shoe soles and floor surfaces], in press.

XP S 73-012 1995, Bottes et chaussures de sécurité. Résistance au glissement sur sols industriels lisses et gras. [Safety boots and shoes—Slip resistance on smooth and greasy industrial floors], (Paris la défense: AFNOR).

CHAPTER 8

Measuring slipperiness—discussions on the state of the art and future research

WEN-RUEY CHANG[1], THEODORE K. COURTNEY[1], RAOUL GRÖNQVIST[2] and MARK REDFERN[3]

[1] Liberty Mutual Research Center for Safety and Health, Hopkinton, MA, 01748, USA

[2] Finnish Institute of Occupational Health (FIOH), Department of Occupational Safety, FIN-00250 Helsinki, Finland

[3] Departments of Otolaryngology and Bioengineering, University of Pittsburgh, Pittsburgh, PA, 15213, USA

An expert workshop symposium of tribologists, biomechanists, clinicians, engineers, epidemiologists and related scientists was held to summarise the state of the art of slipperiness measurement and identify future research needs. Experts were not able to reach consensus on an ideal measurement approach to assess slipperiness. However, experts were able to reach consensus on the necessity of the value of slipperiness measures, outline the criteria for evaluating existing and future measurement systems, and identify key gaps in our knowledge necessitating future research.

1. Introduction

The design of the Liberty Mutual Research Center Hopkinton Conference, 'The Measurement of Slipperiness', from which this book is derived (see Preface), included both the development of state-of-the-art reviews on key issue areas (Chapters 1-7) and a series of discussions among participants within and across the expert working groups who authored the reviews. These discussions attempted to elucidate, individually and collectively, the extent to which any consensus existed in the study of slipperiness measurement and to identify important topics for future research. This concluding chapter presents the results of these discussions.

2. Method

2.1. Individual queries

Individual participants, with the exception of the conference organisers (this chapter's authors), were asked to respond in writing to several questions following the presentation and defense of each of the state of the art reviews:

1. Are we measuring slipperiness adequately?
2. What are the two most crucial research questions in slipperiness measurement?
3. To what extent is the measurement of slipperiness necessary for the prevention of slips and falls?

Participant responses were analysed following the meeting. Attention was given to capturing each participant's statements. Individual responses were then compared and integrated into common themes.

2.2. *Working group queries*

The six expert working groups – concepts and definitions, epidemiology, biomechanics, human-centred, roughness, and friction – held meetings during the conference. During these meetings, each expert working group was asked to respond to the following questions:

1. What are the five most important evaluative criteria to be applied to a slip measurement system?
2. What consensus statement can you make regarding your working group's area of expertise or state of the art as it regards slipperiness measurement?
3. Each working group was asked to bring its responses to the general conference discussion.

2.3. *Conference discussion*

The conference discussion was moderated by one organiser (MR) while detailed discussion notes were developed by another organiser (TC) and projected onto a viewing screen so that discussants could see how their comments were being summarised (and others could take exception if desired) in real time. The conference discussion was also videotaped so that a full record would be available for comparison with the transcripted summary.

Each of the working group questions was addressed sequentially in the discussion. Needs for future research were then discussed collectively. The transcript summary and the videotape were subsequently studied by two organisers (WC, TC) to ensure the capture of critical information. Responses were then aggregated and common themes identified.

3. Results

The results are presented in the order of the questions posed to the participants. In each case, the summary describes the agreements and disagreements among the participants. The final section in the results outlines the top future research needs identified by the working groups and their members.

3.1. *Individual responses*

3.1.1. *Necessity for slipperiness measurement in prevention of slips and falls*

Fourteen of nineteen potential respondents returned a completed question-naire (74% response rate). Responses were grouped into one of three categories based on the intensity of respondent's comments. Three respondents (21%) argued that slipperiness measurements were 'essential' or 'critical'. Six (43%) stated that such measurements were 'necessary'. The remaining five respondents (36%) indicated that slipperiness measurements were at least 'somewhat necessary'. No respondent indicated that these measurements were unnecessary.

3.1.2. *Are we measuring slipperiness adequately?*
Eighty-six per cent of the participants (12) did not feel that slipperiness measurement was adequate. These negative respondents indicated that:

- there was still too little standardisation in methods;
- methods did not yet adequately represent human walking (bio-fidelity);
- the biomechanical data on slipping were too limited;
- an integrative systems approach incorporating surface, footwear, individual, and situational factors was still needed;
- there was no gold standard for device measures in their relationship to actual slipping, permitting too much 'wiggle-room' in the field;
- generalisability in terms of human variability (children, disabled, older workers) was yet to be realised.

Positive respondents acknowledged these limitations but still felt that substantial progress had been made and that a significant body of data was now available.

3.2. *Working group responses*
3.2.1. *Consensus statements*: Only three of the six expert working groups were able to reach any consensus regarding their issue area: epidemiology, roughness, and friction. The epidemiology working group concluded the following:

> *There is persuasive evidence of a substantial contribution of slipperiness to the morbidity and mortality associated with slips and falls. Current surveillance systems are very limited in their ability to adequately define the true extent of slipping as a risk factor. The data available at present are therefore, at best, an underestimate.*

The roughness group concluded that:

> *An understanding of roughness is important to getting consistent friction results due to the influence of roughness on friction.*

Lastly the friction group concluded that:

> *Friction is an important parameter controlling slips and falls, but it is not the only one (slip resistance is a much more global and complicated problem than friction measurement).*

3.2.2. *Most important evaluative criteria*: Group and conference discussions yielded major criteria for evaluating slipperiness measurement systems. Four criteria emerged:

- validity; which includes biomechanical fidelity, tribological fidelity, environmental fidelity, human response fidelity (for human-centred approaches) and flexibility;
- combining measurements with several different methods for assessing slipperiness;
- relating the results of friction measurements to the actual risk involved;
- repeatability and reliability.

3.3. *Future research needs*

Individual and group responses regarding future research needs were pooled, assessed and integrated into related themes. Participants individually and collectively suggested that crucial research needs existed in the following areas:

- Surface aspects
  - improved understanding of friction and wear and surface characteristics at micro-level
  - temporal and spatial variation of slipperiness on a particular surface

- Human sensory and biomechanical aspects
  - more kinematic and kinetic data
  - understanding the what, when, and how aspects of human compensatory responses to apparent or experienced slipping

- Systems approach/modelling
  - relative contributions of sensory systems/balance coordination and surface factors;
  - integration of dynamic body behaviour with surface and footwear characteristics;
  - real situation modelling;
  - systems approach to human variability, holistic-systems approach

- Preventive validity
  - relating slipperiness measurement results to actual injuries
  - testing whether improved frictional characteristics or slip resistance in general reduce actual injury experience
  - better epidemiology tied to surface conditions
  - improved, reliable data sources that can more accurately define the scope of the problem

- Standardisation
  - repeatability/reliability metrics
  - standardisation of test methods (angles, sliding velocities, etc.)
  - differences and correlations among results from different friction measurement devices
  - standard methods of surface preparation (e.g., test sample abrasion procedures, floor surface contamination, etc.)

4. Discussion

4.1. *State of the art*

The results suggest an expert consensus that slipperiness measurements are at least necessary to our understanding of slips and falls. In general slipperiness measures are necessary for estimating slip hazards and to some extent slipping risks, but slipperiness measures may not be necessary or sufficient for estimating the risk of falling and injury (see Chapter 1). For the latter, other factors such as perception of hazards and/or adaptation to a specific risk situation also play a role.

The common interest which drew the participants together and the perhaps subtle and often minority voices which argued that much has been accomplished testify that sufficient intellectual cohesion exists to bring the technical questions into sharper focus. However, the fact that only three of six expert working groups could develop a consensus statement and the generality of those consensus statements that were developed suggests that there remains a healthy diversity of technical perspectives regarding the measurement of slipperiness. In fact the majority of experts believed that slipperiness was not being adequately measured and that substantial gaps exist. These gaps include:

- refining the knowledge in various areas such as heel/floor interactions and behavioural adaptations to slipping;
- integrating the knowledge from the various areas (tribology, gait biomechanics, human situational awareness and proprioception) into a practical systems model of slipping and falling;
- a lack of sufficient standardization in methods and experimental approaches to permit the aggregation and comparison of data from various laboratories and studies.

4.2. *Future research*

4.2.1. *Evaluative criteria*: The evaluative criteria for an ideal slipperiness measurement system reflected many of the concerns raised about 'gaps' by the adequacy question. Validity, in the context argued at the conference, was generally expressed as 'bio-fidelity'. Participants differed in their working definitions of this term. However, by the close of the discussion it was clear that bio-fidelity was meant to include:

- biomechanical fidelity (How closely does the measurement reproduce human locomotion and slipping behaviour?)
- tribological fidelity (physics of interface between footwear and walking surface – film effects, contact area-related issues),
- environmental fidelity (using the actual shoe materials that would typically be used in that environment, typical contaminants) and,
- flexibility to incorporate within and across subject variation for the normal working population and special populations (elders, children, disabled).

Some participants extended the bio-fidelity concept even further into what most considered to be preventive validity—the relationship between the measurement and actual slipping and falling. This was particularly of interest to the friction group whose domain arguably has the most comprehensive data of any of the issue areas. In this case an ideal measurement system would be evaluated as to its predictive or at least correlative relationship to actual slipping and falling. One participant argued that in fact the most one could ask of a measurement method was a relationship to slipping but not to falling since this involved a sequelae of slipping.

The need for multiple measurement approaches or an integrative systems approach was also championed for slipperiness measurement systems. An ideal approach would combine several dimensions of the slipperiness question and be able to incorporate variability in various risk situations (such as walking on the level and inclined surfaces, carrying loads while walking, walking on stairs, pushing and pulling, etc.). Such an approach might well involve both surface and human-centred evaluations.

The ability of a measurement system to consistently report the same value(s) given the same testing conditions was also raised as a key factor. A measurement system should be operator independent. Here there were several participants who pointed out that reliability/reproducibility may be difficult with surface measures since actual surface conditions are rarely uniform and can introduce a variation which would not be attributable to the measurement system itself.

Additionally participants raised several issues related to the actual useability of measurement systems (how user-friendly they are). A system should be reasonably portable if it is to be used in actual application environments. It should be capable of being operated with minimum stress to the operator and present information in such a way that user errors are minimised.

It was widely acknowledged that no one device in existence could adequately address all of these parameters. In fact several argued that the criteria would be very difficult to meet for a field device that could be operated without extensive training or technical knowledge. However, these criteria were seen as presenting the conceptual framework for evaluating a system rather than the absolute metrics themselves.

4.2.2. *Needs for future research*:   The conference participants outlined a number of specific research needs (summarised in section 3.3) from basic materials testing methodologies to human factors. In a number of cases, these research topics harmonise with those identified in Chapters 1–7. In this section, each topic will be referenced to the chapter or chapters which contain relevant related material.

Like any measurement system, the measurement of slipperiness needs to be accurate and repeatable. However, the measurement must relate to the potential for falling during various human endeavours such as walking, turning, pushing, pulling, or lifting. Thus, the challenge for the research community is to develop an accurate measurement that is relevant to the outcome (see Chapter 1). This will include learning more about the basic interactions of the shoe-floor interface characteristics and the influence of contaminants on these interactions. In addition, measurement of these interactions must be considered in the 'physiological range'. Finally, the reaction of the human postural control system to slipping is an important factor in determining whether a particular environment will result in a fall.

There remain a number of interesting research questions related to surface factors (see Chapters 5–7). The contribution of the actual friction in a given physical location versus the magnitude of the difference between that location and contiguous locations has not been answered. It could be that slipperiness has more to do with *abrupt transitions* in friction values than with very high or very low values. In addition to spatial variability, there is still much to understand about friction variability over time. For example, how do contaminants or surfaces change in their tribological characteristics due to evaporation, heating, cooling, or oxidation? In addition the microscopic factors (friction mechanisms, lubrication dynamics, and wear) and their interactions at the shoe–floor interface are not yet well understood. For example, what influence do squeeze-film and lubrication effects have on friction variability?

There was consensus from the conference that the best approach to slipperiness measurement was to measure under 'biofidelic' conditions (see especially Chapters 3, 4, and 6). In other words, measure slipperiness under conditions that mimicked the actual forces and motions of the shoe–floor interface during human tasks such as walking. There was not consensus on exactly how this should be accomplished. Some participants felt that friction measurements that used forces and motions comparable to walking were the best approach. Others felt that using human-centred measures was preferable due to the complexity of the tasks. Further research in this area is needed to examine the utility of both approaches given the requirements of accuracy and repeatability.

Another identified area of research was that of preventive validitiy (see especially Chapters 2 and 4). The fundamental question is: can the measurement of slipperiness predict falling? Unless this is true, a measure, even if repeatable and accurate, will be of little use. There is a great need for studies of the preventive validity of the measures. This is not an easy task; however, further work needs to be performed in this area. Ideally, data from these studies would be able to be used to evaluate current slip measurement systems and those developed in the future.

Much still remains to be learned regarding the interaction of human postural control and slipperiness (see Chapters 3 and 4). Postural responses allow the maintanance of balance under a variety of conditions, and the postural control system adapts to different environments and situational requirements. There needs to be greater understanding of the interaction of slipperiness and postural responses in varying situations. Slip resistance requirements to avoid falling and perform tasks (e.g. pushing, pulling) may be a function of postural control. In addition, the postural control system is affected by factors such as age, musculoskeletal health, sensory health, cognitive abilities, and many types of disabilities.

Finally, the question of standardisation of slipperiness measurements was raised (Chapters 6 and 7). Clearly, it would be of benefit to establish standards in measurement that can be applied across national and international environments. The aggregation of such standardised data would enable a more rapid advance in the investigation of the global issues in slipping and falling. Given the diversity of measurement devices already developed, it is likely that initial inroads can be made here in terms of the standardisation of test parameters such as those indicated in Chapter 6. Standardisation to a single device design remains a potentially attractive yet elusive ideal reflected in the consensus of the lack of any current 'gold standard'.

5. Conclusion

Clearly, important strides have been made in the measurement of slipperiness over the past 30 years. However, the morbidity and mortality related to slipping and falling remain significant. By examining the current technical opinions of the majority of expert conference participants individually and collectively, this chapter has sought to contextualise the state of the art for the reader who has already progressed through the detailed content of the first seven chapters.

Experts could not reach a clear consensus on an ideal measurement approach to assess slipperiness. However, experts were able to reach a consensus on the value of slipperiness measures, outline the criteria for evaluating existing and future measurement systems, and identify key gaps in our knowledge necessitating future research. The challenge remains ahead of each of us who would better understand slipping to devote ourselves to reducing these gaps and thereby to reducing the burden of slipping in our global society.

Author Index

Aalto, H., 69, 70, 71, 97
Abeysekera, J., 2, 4–6, 10, 12, 14, 61, 63, 67, 80, 92, 93, 120, 133
Aghazadeh, F., 15
Ahagon, A., 128, 132
Akkok, M., 128, 132
Albarede, J. L., 63
Alexander, N.B., 69, 93
Allum, J.H.J., 70, 93
Almshult, B., 127, 132
American Society for Testing and Materials (ASTM), 15, 63–65, 93, 97, 134, 137, 138, 142, 144, 146, 147, 161–163
Amontons, M., 38, 62, 102
Amos, W.B., 109, 117
Anderson, C., 15, 97, 134, 161, 162
Andersson, G.B.J., 52, 62
Andersson, R., 2, 9, 13, 21, 33, 38
Andersson, S.I., 79, 94
Andres, R.O., 41, 64, 86, 93, 121, 132, 134, 158–161
Andriacchi, T.P., 52, 62
Archea, J., 98
Aromaa, A., 69, 96
Association Francaise De Normalisation (AFNOR), 142, 146, 163
Ayers, I., 2, 3, 15, 18, 31, 34, 35, 72, 96

Bak, S., 72, 74, 96
Ballance, P.E., 136, 161
Banks, S.A., 41, 62
Barnes, P.M, 17, 35
Barquins, M., 124, 132
Bejan, A., 128, 132
Bentley, T.A., 31, 33, 47, 63
Berg, K.O., 79, 93
Berthe, D, 132
Berthier, Y., 127, 133
Bhattacharya, A., 15, 68, 78, 93, 97, 134
Bhushan, B., 108, 115
Bidanda, B., 39, 42, 64, 148, 149, 163
Black, O.F., 97

Blanchfield, L.P, 2, 3, 15
Bloem, B.R., 70, 93
Blom, K., 28, 33
Bloswick, D.S., 2, 3, 5, 15, 41, 52, 64
Bohannon, R.W., 79, 93
Bonefeld, X., 85, 86, 98
Bongers, P.M., 3, 14
Bonnard, M., 70, 94
Boone, T.H., 102, 117
Bouter, L.M., 3, 14
Bowden, F.P., 102, 115, 127, 132
Brady, R.A., 60, 62, 68, 76, 83, 90, 93
Brakenhoff, G.T, 109, 117
Brand, R.J., 32, 34
Brault, J.R, 144, 154, 163
Braun, R., 123, 132, 159, 161
Brechter, J.H., 52, 53, 65
Bring, C, 85, 93
British Standard Institution, 103, 115
Bruce, M., 2, 3, 15, 18, 31, 35, 72, 78, 93, 96, 102, 111, 116
Brungraber, R.J., 2, 4, 6–8, 11–14, 61, 63, 68, 85, 93, 110, 119, 131, 132, 135, 136, 138, 142, 156, 157, 159, 161
Buczek, F.L., 41, 62, 63, 87, 93
Bulajic-Kopiar, M., 74, 93
Bunterngchit, Y., 2, 4, 8, 11, 13, 61, 63, 101, 109, 110, 117, 121, 133, 146, 148, 149, 163
Burdorf, A., 9, 13
Burns, P.C., 34, 78, 94

Calabrese, S.J., 128, 132
Cappozzo, A., 4, 13, 130, 132
Capra, R., 71, 97
Carlini, A.R., 109, 117
Carlsöö, S., 4, 7, 13, 85, 86, 93
Carpenter, M.G., 70, 93
Cavanagh, P.R., 41, 63, 86, 87, 93, 94
Chaffin, D.B., 10, 14, 158, 159, 160, 161
Cham, R., 2, 4–7, 10, 12, 13, 15, 37, 39–41, 43–45, 47, 48, 52, 54–59, 63, 68, 71, 73, 76, 77, 85, 86, 93, 97, 120, 134

Subject Index

Printed and bound by CPI Group (UK) Ltd, Croydon, CR0 4YY

23/10/2024

01778248-0004